EMERGENCE

EMERGENCE

A Philosophical Account

Paul Humphreys

OXFORD
UNIVERSITY PRESS

OXFORD
UNIVERSITY PRESS

Oxford University Press is a department of the University of Oxford. It furthers the University's objective of excellence in research, scholarship, and education by publishing worldwide. Oxford is a registered trade mark of Oxford University Press in the UK and certain other countries.

Published in the United States of America by Oxford University Press
198 Madison Avenue, New York, NY 10016, United States of America.

CIP data is on file at the Library of Congress

ISBN 978-0-19-062032-5

1 3 5 7 9 8 6 4 2
Printed by Sheridan Books, Inc., United States of America

To Emily Victoria and Alexandra Elizabeth

CONTENTS

ACKNOWLEDGMENTS

I am profoundly grateful to the many philosophers and scientists who have listened to preliminary versions of this material, some of it now abandoned, engaged in vigorous discussions about a subject on which little agreement exists, and provided valuable input over the years. I cannot possible acknowledge them all, but especially helpful were Anouk Barberousse, Bob Batterman, Mark Bedau, Michel Bitbol, Georges Chapouthier, Dong Chunyu, Jean-Paul Delahaye, Fan Dongping, Jacques Dubucs, Nina Emery, Sarah Franceschelli, Serge Galam, Alexandre Guay, Zhang Huaxia, Philippe Huneman, Andreas Hutteman, Cyrille Imbert, Fred Kronz, Yinyin Kuang, Johannes Lenhard, Christophe Malaterre, Margaret Morrison, Olivier Sartenaer, Eric Scerri, Jim Simmonds, John Symons, Jessica Wilson, and Bill Wimsatt. Development of the ideas in this book would not have been possible without financial support from an ACLS fellowship, a foreign research fellowship at the IHPST in Paris from the Centre National de la Recherche Scientifique, an NSF Scholars Award, and a Sesquicentennial fellowship from the University of Virginia. And, as always, thanks to Diane.

PREAMBLE

After a period of quiescence lasting around fifty years, the topic of emergence is once again a flourishing field. There are many reasons for this, but three of the important influences have been the discovery in various sciences of phenomena that look like plausible candidates for emergence, the rise of techniques within complexity theory that have provided mathematical tools for representing emergent phenomena, and the philosophical development of conceptual analyses that advance our understanding of what emergence might be. The situation is thus very different than it was in the early 1930s, when interest in emergence declined. Then, the suggested candidates for emergent phenomena were things that were not well understood, such as life, consciousness, and reflective mentality. As the mysteries surrounding those candidates diminished and our understanding of the processes underlying them improved, emergentist approaches were abandoned. In the interim, new and better understood candidates for emergence have appeared. Many of the candidates, including self-organizing systems such as slime molds, phase transitions leading to the appearance of ferromagnetism, and the flocking behavior of birds have well-understood models, and the obscurity that shrouded the earlier possibilities for emergence has disappeared. What is often left implicit in the scientific literature is why examples such as these count as emergent. One goal of this book is to explicitly identify the criteria for emergence.

THE PROJECT

That goal is pursued here with the aim of providing a conceptual framework within which to locate scientifically driven examples. There are three projects involved: the task of understanding what emergence is; the task of constructing consistent theories of emergence; and the task of establishing whether the world contains any examples of emergent phenomena. From the point of view of philosophy alone, one might be content with only the first two of these projects. Indeed, at various points I shall use simple artificial examples to convey the core conceptual features of a particular approach to emergence. But the desire to know whether real examples of emergence exist is entirely reasonable. It would be a striking feature of our universe if it did not contain a single example of an emergent phenomenon, so striking that an explanation would be called for. In addition, merely conceived possibilities lack the detail of scientific examples that leads to an improved understanding of how emergence can occur. One of the most serious flaws in the early investigations of emergence was that the examples of emergence put forward, such as chemical properties and life, seemed plausible only because the state of knowledge at the time precluded alternative, non-emergentist explanations of the examples. A related situation exists today. When phenomena such as consciousness and perceptual qualia are proposed as examples of emergence (not to mention even more esoteric cases such as souls), the current state of scientific knowledge in neuroscience is insufficient to assess whether these are genuine cases of emergence. They tend to qualify by default, because there is no generally accepted and transparent theory of these phenomena. The explanatory gap results from a knowledge gap that might be bridged in the indefinite future. This kind of situation was considered acceptable when emergent phenomena were taken to be inexplicable, but I do not adopt that defeatist view here.

TYPES OF EMERGENCE

The type of emergence in which I am primarily interested is ontological emergence. This asserts that genuinely novel objects and properties emerge even within the domain of physics, and it rejects the idea that only the level of fundamental physics is real. The consequences of acknowledging the existence of ontological emergence are far-reaching, not only for physicalism but also for other deeply entrenched philosophical positions. Despite the importance of ontological emergence, it is crucial to keep in mind that there are multiple concepts of emergence, each of which legitimately captures some types of emergent phenomena, but none of which provides a complete account of emergence. There does not exist, and I believe there will not exist in the near future, a single, overarching theoretical framework for the topic.[1]

There are two primary reasons for this. The first is that the concept of emergence is not rooted within a single science, in contrast to such concepts as gravitation and heredity. Emergence is claimed to occur in physics, in chemistry, in biology, in complexity theory, in economics, and in a host of other domains. So, no single scientific domain can impose itself on the concept in a way that general relativity can impose its theoretical framework on the concept of gravity. This transdisciplinary aspect does not by itself prevent a unified treatment, for we have the example of causation as a multidisciplinary success story.[2]

In noting this, we encounter a second reason for the heterogeneity of approaches to emergence. It is that we do not have a firm pretheoretical grasp of emergence in the way that we do with something like causation. Compare what happens if you survey individuals about what they consider to be clear examples of emergence and what they consider to be clear cases of causation. Almost every linguistically competent human being who allows that causation exists would agree that certain core examples, such as a rock shattering a window, are genuine cases of causation. This is not true for examples of emergence, even among those who agree that emergence exists. It is not simply that you will get different examples, but also that researchers in one area will

1. This view is shared by Holland (1998, 3): "it is unlikely that a topic as complicated as emergence will submit meekly to a concise definition."
2. Even though there is evidence that we have more than one concept of causation, each such concept can be applied across more than one type of subject matter.

vigorously contest the validity of examples put forward by workers in another area.[3]

Because of this variety of approaches, potential counter-examples to any given account of emergence need to be treated with care. An inferential account of emergence need not satisfy criteria that apply to an ontological account of emergence, nor should diachronic approaches to emergence be forced to accept conditions that properly apply only to synchronic accounts. One of the obstacles to progress in this area, and the source of many unproductive disputes, has been the view that there is one correct definition of emergence. In an intellectually ordered world, we would stop using the word "emergence" and use more precisely defined terms such as "self-organizing system" and "computational incompressibility," together with features that are characteristic of, but neither necessary nor sufficient for, emergence. Doing so would go a long way toward preventing confusion. Yet, the word has an irresistible attraction and I do not have the power to ban it. So, with the caveats already mentioned, and with modifiers attached to capture different types of emergence (without the assumption that they are all species of an underlying genus), I shall retain the term as a convenience.

DIACHRONIC AND SYNCHRONIC EMERGENCE

Running through these various approaches to emergence is a very basic distinction between synchronic emergence and diachronic emergence. Synchronic emergence considers the emergent phenomenon to exist simultaneously with the features from which it emerges. A commonly cited type of example is when a pattern develops in a system, and that pattern is essentially global in the sense of belonging to the system as a whole and being irreducible to properties of the system's components. In

3. A survey of positions on emergence that I conducted with astronomers and philosophers showed clear differences between the two fields as to what were considered to be examples of emergence. The sample size was too small to warrant any general conclusions, but the survey did confirm the experiences I have had over many years of discussing emergence with a broad spectrum of scientists, philosophers, and engineers. There are fundamental and irreconcilable differences between individuals' attitudes toward emergence.

contrast, diachronic emergence takes emergent phenomena to develop over time from preexisting states of the system. An example might be the emergence of living organisms as a result of self-organizing processes in an artificial-life experiment. There was a tradition of emergence known as evolutionary emergence that was popular in the 1920s, but much of the current philosophical literature on emergence is exclusively about synchronic emergence. This not only results in an unnecessarily narrow focus, but also makes it impossible for diachronic forms of emergence to satisfy the synchronic conditions. A principal theme of this book is that in many cases it is diachronic processes that produce emergence and that the philosophical emphasis on synchronic emergence is a distraction.

CONSTRAINTS

One constituency that has been interested in emergence will have its concerns addressed only occasionally. Within philosophy, a great deal of attention has been paid to the question of whether or not mental states can be reduced to brain states. It is worth keeping in mind that the mind/brain reduction question is not representative of other cases of reduction or nonreduction. Mental states have features such as intentionality, the ability to be about states of affairs that may not exist, such as "Allie believes that Harry Potter is real." This feature is peculiar to the mental realm and areas derivative from it, including some social systems in which collective intentionality plays a central role.[4] Moreover, despite significant advances in the area, the current state of knowledge in psychology and neuroscience remains underdeveloped compared to physics and chemistry. An emphasis on the mental can, therefore, be deeply misleading when discussing emergence, because it reinforces the view that cases of emergence are mysterious and rare. There are better understood cases of emergence in the physical sciences that are more representative of emergent phenomena. It is for this reason that I have in

4. There may be primitive forms of intentionality in simple biological systems, but the attribution may also be a case of secondary intentionality imposed by us to help interpret the behavior.

general stayed away from examples drawn from psychology, and I have little specific to say about the philosophy of mind, although I shall discuss some issues that were originally motivated by topics in that area.

Accompanying this restriction is the fact that physicalism, considered as one half of a physical versus mental taxonomy, is too crude a classification to be interesting, and a position that indiscriminately contrasts physical phenomena to all others will look in the wrong place for emergent entities. We shall see that emergence is not uncommon in physics and chemistry. If it turns out that the more complex sciences such as neuroscience also contain numerous emergent phenomena, so much the better. For the record, I expect that many of the most interesting features of the mind will turn out to be associated with emergent phenomena.

1

Basic Features of Emergence

> Coal and diamonds, sand and computer chips, cancer and health tissue: throughout history, variations in the arrangement of atoms have distinguished the cheap from the cherished, the diseased from the healthy. Arranged one way, atoms make up soil, air and water; arranged another, they make up ripe strawberries. Arranged one way, they make up homes and fresh air; arranged another, they make up ash and smoke.
>
> Drexler 1986, 3

1.1 INTRODUCTION

A persistent view about the nature of reality takes the following form: The world is nothing but spatiotemporal arrangements of fundamental physical objects and properties. You and I, rocks and galaxies, toads and scrambled eggs are just processes, the successive states of which are spatial arrangements of elementary physical objects. These elementary objects, arranged in different configurations, account for all the astonishing variety that we encounter in our day-to-day and scientific lives. There are both horizontal and vertical structures in this picture. The horizontal structure is provided by processes within which earlier states of systems are causally responsible for later states of the same or other systems. The vertical structure consists in a layered ontology, with the objects and properties of fundamental physics at the foundations, the objects and properties of nuclear and atomic physics above those, the subject matters of chemistry, biochemistry, and cellular biology above those, followed by the objects studied by the various psychological and social sciences.

This, in its crudest form, is the *generative atomism* version of the physicalist picture—everything in the world is generated from combinations of elementary physical objects and their properties.

Generative atomism has, in the form just given, limited scope at best, but as our chapter epigraph indicates, it is frequently used as a rhetorical device within science. It has the considerable advantage that it lends itself to representations in mathematics and logic that are, in general, tractable and transparent—features that should not be underestimated. The computational models that have become an increasingly dominant method in many sciences rely heavily on an approach founded upon generative atomism. Agent-based models in economics and computational chemistry, although heavily laden with approximations and idealizations, are driven by methods that parallel the ontology of generative atomism. In philosophy, the tradition of logical analysis that has dominated Anglo-American philosophy at least since Bertrand Russell's 1918 lectures on logical atomism (Russell 1918) remains influential, a notable example being the continued use of mereology within metaphysics.

Generative atomism has an ontological and a methodological aspect. In its ontological form, the position is that non-atomic entities are ultimately composed of atomic entities, with structure provided by relations appropriate to the type of entity involved. In the methodological version, not only must we show that the composite entities can be decomposed into sets of lower level entities, but we must also show, starting with those lower level entities, that the higher level entity can be generated from those entities. As a practical matter, the difficulties of explicitly reconstructing the higher levels from the lower levels are, in many cases, so severe that a full reconstruction of the higher levels from the lower levels is not feasible and a new theoretical apparatus must be introduced in order for predictions and descriptions to be effectively carried out. Despite those difficulties, generative atomism has undoubtedly been successful in some limited domains. In others, the situation is more complicated. The properties of small atoms such as helium have, according to legend, been predicted using features that are dependent only upon spatial arrangements of various fermions and bosons. We shall see later that these claims are not quite as clear-cut as they might seem, but for the moment we can grant their approximate truth. In more complex cases, the methods of generative atomism cannot be carried out in practice, and a less ambitious program than the combinatorialist's is substituted. In this alternative approach, the lower levels, and perhaps the lowest level itself, are said to determine everything else in the following sense: once the elements and structure of the lowest level are present, the objects, the structure, and the laws of the remaining levels are fixed by the individuals and the properties of the lower levels. The fundamental laws of the universe govern the spatiotemporal arrangements at the lowest levels, and everything else that goes on is determined by those arrangements.

Of course, this is a deliberately simplified account of most systems. It relies on accepting a physicalist ontology and accepting that spatiotemporal relations are fundamental. These assumptions are controversial, with field theories, pure structures, and information serving as possible alternative ontologies and higher dimensional spaces serving as structures that replace spacetime, but the position just described is one to which various accounts of emergence are often unfavorably compared.

The combinatorial methods behind this picture have served science and philosophy well for many centuries, and they remain highly effective tools in many areas. Yet, we have left behind the era in which generative atomism and its more sophisticated cousins are tenable as a default method for representing the world. Their successes were due to a selection effect; the kinds of systems that lend themselves to their successful use are relatively simple and have special features that are absent in many more complicated systems. One goal of this book is to show why even considerably more sophisticated versions of the positions just sketched are not capable of providing correct representations of many physical systems. The primary culprit is an excessive emphasis on synchronic relations—the vertical structure just mentioned. The horizontal processes—the diachronic aspects—are more important to emergence than is generally conceded.

1.1.1 The General Approach

The power and historical successes of generative atomism are ample reasons for not abandoning the approach lightly. Indeed, one of the running themes of this book is that the willingness to classify a phenomenon as emergent is always a result of some version of generative atomism or its close relatives' failing. That is, an approach based on a compositional ontology or representation is used as the basis of the research, the evidence shows that the desired goal cannot be obtained, the reason for the failure is isolated, and that reason then becomes the basis for an account of emergence. Generative atomism can fail in a number of different ways and because there are both ontological and inferential versions of generative atomism, different, albeit generically related, accounts of emergence can be given. To take just one example, C. D. Broad's inferential approach to emergence, which has been adopted and adapted by many contemporary philosophers, is a particularly clear case of starting from the failure of generative atomism. Broad took what he called Pure Mechanism as an ideal form of a compositional ontology and described how emergent phenomena would appear within systems that deviated from that ideal. Here is how he presented it:

> I come now to the second type of theory which professes to explain, wholly or partly, the differences of behaviour between different things.... On the first form of the theory the characteristic behaviour of the whole *could* not, even in theory, be deduced from the most complete knowledge of the behavior of its components, taken separately or in other combinations, and of their proportions and arrangements in this whole. (Broad 1925, 58–59)[1]

More generally, the essential idea is that with generative atomism, facts about composite entities are implicitly included in facts about the atoms, together with facts about the rules of composition. As a result,

1. As one example of Broad's legacy we have: "For [a system property to be reducible], two conditions must be fulfilled: The first is that from the behavior of the system's parts alone it must follow that the system has some property P. The second condition demands that the behavior the system's parts show when they are part of the system follows from the behavior they show in isolation or in simpler systems than the system in question" (Stephan 1999, 51).

there is nothing essentially new about the composite entities. This is not so when emergence occurs—some entity falls outside the domain of those produced by the generating processes, whether these processes be ontological, predictive, or conceptual. This is the origin of two characteristic features of emergence. The first is that there is some aspect of the emergent entity that is novel. The second is that because at least one fact about the novel aspect is not implicitly contained in the atoms plus the rules, it must be a holistic feature of some non-atomic entity. The sketch just given is general, and in the particular example the determination relation is an inferential relation. In the ontological case, the properties of the atomic constituents together with the laws governing those properties do not ontologically determine the properties of the non-atomic entity, whether synchronically or diachronically. There can be different reasons for this failure of determination. In some cases, the non-atomic entity has as parts entities that are not themselves composites of atoms; in others, the new object is not composite; in still other cases, additional laws are needed to determine the properties of the non-atomic entity.

1.2 A GENERAL STRATEGY

There was no entry on emergence in the 1967 edition of the *Encyclopedia of Philosophy* (Edwards 1967). Given the historical context, that was not surprising, for the omission accurately reflected a widely held view that emergence was a disreputable subject, one that had been killed off within science by advances in quantum chemistry and molecular biology and within philosophy by a hard-nosed approach to theory reduction. The theory of the chemical bond, based on quantum theory, had explained why the properties of water were different from those of its constituents, hydrogen and oxygen; the mysteries of life were being decoded after the discovery of the structure of DNA; and physicalism was beginning its rise to dominance. The *Encyclopedia* did provide an article on Emergent Evolutionism, a largely historical essay that serves to remind us that during the period 1915 to 1935, diachronic approaches to emergence—those that emphasize the development of novel features across time—played an important role in philosophical discussions of the topic.[2]

This lack of interest in emergence reflected the contemporary attitude. The stark contrast between the speculative content of the available emergentist positions and the clarity of the generative atomism projects made the choice between the two easy during much of the twentieth century. In recent years that has changed. Work in complex systems theory has led to sophisticated mathematical theories of diachronically emergent phenomena. Progress in condensed-matter physics has cut into the nomological hegemony of high-energy physics. The discovery of computationally incompressible processes has brought a healthy dose of realism to previously abstract approaches to axiomatically formulated theories. In philosophy, holistic theories of meaning and evidence have made holism more familiar, perhaps even acceptable, and anti-reductionist positions have been articulated that allow for the existence of autonomous layers of properties.

1.2.1 Method

Still, the taint of mysticism clings to emergentism. Although there is no question that some of the literature on emergence has been less than clear, this is hardly special to the topic of emergence. Indeed, one of the curious things about emergence is that we have a reasonably good intuitive grasp of what it might be. For example, when two companies, MegaCorps and ZirCon, merge to form GigaCon, we can no longer say that MegaCorps is located here and ZirCon is located there; we can sometimes, when the merger is a success, appropriately say that the new company emerged from the former organizations, that it is different from the combination of its former parts, that its organizational structure is autonomous from those of its predecessors, that the new company has novel features not present in either predecessor, and so on. This intuitive grasp may well be illusory, and the example relies on comparing

2. For a history of the Emergent Evolution movement, see Blitz (1992), and for a then-contemporary assessment of emergence, see E. Nagel (1961, chap. 11) and Hempel and Oppenheim (1948, 146–52).

the functioning of the combined and separate organizations, but its prima facie appeal as an example of emergence indicates that emergence is not beyond the reach of careful articulation.

So how should we approach this tantalizing subject? Because of the disagreement over what counts as a core example of emergence, noted in the preamble, we cannot start with examples and use an abstractive bottom-up process to construct a general theory of emergence. Instead, we must start with general principles about particular approaches to emergence, coupled with a feature that seems to accompany all the contexts within which claims of emergence are made. The feature is this: emergence is taken seriously when a successful generative method fails to accommodate a phenomenon that we have antecedent reasons for believing should fall under the scope of that method. Generative atomism fails at the quantum level, and emergence talk begins. Supervenience fails for certain aspects of conscious experience, and the possibility of emergence is raised. The construction of macroscopic descriptions from our theories of microprocesses so greatly exceeds even idealized computational resources that novel concepts describing emergent phenomena must be introduced. The existence of undecidable propositions in physics indicates that our predictive techniques have essential limits, and the term "emergence" is introduced. None of this is definitive, but it does suggest a promising way to proceed. I shall identify six core types of emergence, each of which is associated with the breakdown of a powerful generative method. Each failure satisfies a cluster concept containing features that are traditionally associated with emergence, thereby warranting the use of the term "emergent."

These generative failures can affect synchronic relations or diachronic processes, whether we are dealing with ontological or with predictive approaches. As we have already mentioned, an unfortunate feature of the recent revival of interest in emergence is the fact that diachronic concepts of emergence have largely been neglected in favor of synchronic analyses of emergence—those that emphasize novel "higher level" objects or properties that coexist with objects or properties that are at some "lower level."[3] Perhaps as a result, contemporary

3. Exceptions to this emphasis on synchronic emergence include Bedau (1997, 2002), Humphreys (1997a), O'Connor and Wong (2005), and Rueger (2000a, 2000b).

philosophical research on emergence has tended to emphasize two features. The first is a metaphysics of levels, the position that there is an ontology that is divided into an ordered hierarchy of levels, with the more fundamental entities occurring toward the bottom of that hierarchy. The second feature is a widely held view that a collection of philosophical tools including supervenience relations, functionalized properties, realizations, second-order properties, grounding, and the like can adequately represent most, and perhaps all, cases of synchronically emergent phenomena and hence most, if not all, cases of emergence simpliciter.[4] I shall argue here that these synchronic approaches have misrepresented what lies at the core of emergence, and that they should be deemphasized in favor of diachronic approaches that give priority to the temporal evolution of the system. The traditional contrast between emergence and reduction, with its emphasis on synchronic reductions, fails to capture an important class of diachronically emergent phenomena, and we should not insist that examples of emergence are always a result of failures of synchronic reduction.

1.2.2 Models and Reality

Because of their importance, it is worth dwelling on the differences between emergent phenomena that occur in models and emergent phenomena that occur in natural systems. We can use examples and arguments based on models to demonstrate the consistency of an account of emergence, whereas we need evidence of the truth of the models to argue that instances of that kind of emergence exist as features of our world. One advantage of using models as a vehicle for discussion is that we know explicitly what constitutes the structure and content of the model. As a special case of this, model systems need to consider nothing more fine-grained than the fundamental entities of the model, unlike physical systems within which there is always the possibility of another level existing below the one currently being considered. One particular class of models, computational models, has a special status. These

4. This apparatus is employed in, among others, van Cleve (1990), McLaughlin (1997), and Kim (1999). Kim's position is based on approaches found in Levine (1993).

are run on real computational devices, and those runs and their outputs form part of the world's ontology. A pattern emerging as the output of a computation is as real as a similar pattern appearing in a biological population. But the processes that give rise to the two patterns will ordinarily be quite different—computational in the first instance, biological in the second.

Models of such phenomena as ferromagnetism, superconductivity, and superfluidity are, like all models, idealized in certain ways. Although there are detailed consequences of these models that fit well with experimental data, the internal structures of the models themselves do not fully represent the structures of the real systems, and in many cases they are known to be gross oversimplifications of the real systems' structures. This is a common feature of models, and because of it we must be very careful indeed when extending claims about the existence of emergence in models to claims about its existence in the systems being modeled. The best we can do is to suggest that in worlds that resemble ours in certain important respects, and which behave empirically like ours in ways captured by those models, something much like emergence can occur.[5] This makes many claims about ontological emergence inescapably tentative.

Some elementary models will play an important role in this book. They will illustrate in what I hope is a clear fashion how a particular kind of emergence can occur. These models, despite their simplicity, have a significant advantage over most examples drawn from science because we are freed from concerns about their truth and whether the model will be overthrown in the future by the discovery of currently unknown empirical facts. Having shown that ontologically emergent phenomena are logically possible by constructing consistent models that have

5. One virtue of using models is that we avoid relying on merely conceiving of something as emergent and thereby also avoid the issue of whether conceivability entails logical possibility. For example, the British Emergentists thought that the way in which they conceived of living entities immediately gave them an example of an emergent phenomenon. Since many contemporary biologists hold that the property of being alive is not emergent, whatever it was that the British Emergentists were conceiving of was, at best, confused. For a contemporary discussion of the conceivability and possibility issue, see Gendler and Hawthorne (2002).

emergent features, we must then address a more difficult task: showing that emergent phenomena are nomologically possible, or that emergent phenomena are consistent with the currently known laws of nature. To carry out this project it is still not necessary to demonstrate that actual examples of emergence exist.

This part of the project has precedents. Physics has investigated uninstantiated properties—for example, properties that are attached to heavy atoms that are not naturally instanced in our world. In philosophy, there has been a traditional separation between the task of capturing what counts as a causal relation and the task of discovering which causal relations are instantiated in our world. However, we cannot completely separate the tasks of providing a consistent account of what emergence might be and of demonstrating that there are examples of emergence in the world. Within the philosophy of science, we are less free than are some other areas of philosophy to propose concepts that have an empty extension. Within epistemology, for example, the task of providing a definition of the predicate "omniscience," can proceed quite happily even if it turns out that it applies to nothing.

To show that a phenomenon is nomologically possible, it is legitimate to consider examples in which idealizations, simplifications, and approximations are used that are consistent with the laws of nature. In contrast, it is impermissible to use idealizations that are nomologically impossible, even though such idealizations are used in physics and other sciences. As long as the idealizations and other refinements are not inconsistent with the application conditions of a law, that law applies as much to idealized cases as it does to non-idealized cases. For example, in mechanics, the fundamental form of the dynamics is very general—it applies to a Hamiltonian with severe idealizations just as much as to a completely intractable but realistic Hamiltonian. In contrast, if idealizations that are nomologically impossible are used to construct an example, this suggests that the existence proof is merely a mathematical result.

An example of the differences between these approaches to emergence is a group of theoretical models of sand piles that use self-organized criticality to generate a power-law distribution for the size of avalanches on the surface of the pile (Bak, Tang, and Wiesenfeld 1987).

These examples have been cited as a kind of emergent phenomenon exhibited by a physical system, and the theoretical models are nomologically possible. However, experiments on real sand piles showed that they did not exhibit the power-law behavior displayed in the models, and so those do not count as real examples of emergence. (See Jaeger, Liu, and Nagel 1989; S. Nagel 1992; Jensen 1998.) However, finite-size scaling phenomena, which are characteristic of self-organized criticality, have been found on the surfaces of three-dimensional piles of long-grain rice (Aegerter, Günther, and Wijngaarden 2003).

1.3 GENERATIVE ATOMISM

With these methodological preliminaries addressed, we can begin with the simplest form of the generative approach, leaving more complicated forms until later. A scientific version of generative atomism, in a somewhat simplified form, has the following features. There exists a collection of fundamental facts, usually regarded as relatively small in number, and often conceived of as being about fundamental physics, that describe the properties of the basic constituents of nature. These fundamental facts are of four kinds:

(a) Facts about the fundamental kinds of entities, preferably in the form of an explicit list with an associated taxonomy.

(b) Facts about the permanent properties of these fundamental constituents, such as their charge and mass, including specific quantitative values for these properties. (If these values, such as the value of the charge, are defining features of the fundamental objects, then (a) and (b) can be collapsed into a single category.)

(c) Facts about what transitory properties the fundamental entities can have, such as the direction of spin, position, momentum, and so on. For example, the value of spin for fermions would be required in a specification of (b), but whether it was up or down for a particular particle would be included in category (c).

(d) Facts about the laws that govern the distribution and dynamics of the fundamental entities and their properties.

As just described, generative atomism is not a fully adequate approach for physically fundamental facts. The method is general, however, and we should not be fixated on fundamental physics for our ontology, lest we be biased against the possibility of emergence elsewhere. Within methodological individualism in the social sciences, the fundamental entities are individual humans, the fundamental facts might include their rationality as a competence (rather than as a performance), the transitory properties might include cooperating or not cooperating with other agents, and the psychological laws would include the employment of decision and game theoretic principles.

Generative atomism in this basic form has both a synthetic and an analytic component. The synthetic component says (1) that there is a collection of elementary entities from which all other legitimate objects in the domain are constructed, (2) there is a fixed set of rules that govern the construction process, and (3) as a consequence of (1) and (2), all entities are either atoms or are composed of atoms.[6] The analytic component asserts that any non-atomic object can be uniquely decomposed into its atomic components using an explicitly formulated set of decomposition rules. It is possible for the analytic aspect of generative atomism to be satisfied by a system while violating the synthetic aspect. One example is the fact that the marginals of a joint probability distribution are uniquely determined by the joint distribution, whereas the marginal distributions, without further information about the relations between the random variables, do not determine the form of the joint distribution. (For more discussion of this point, see section 2.3.2) It is also possible to subscribe to the synthetic component of generative atomism while rejecting the analytic doctrine because the synthetic rules are

6. For one description of this process, see Emmeche, Køppe, and Stjernfelt (2000, 16), where they claim that this approach is involved in both "constitutive reductionism" and "constitutive irreductionism." The process that P. W. Anderson calls "constructionism" (Anderson 1972; for more on this, see section 5.2) is in the synthetic tradition of generative atomism. O'Connor and Wong (2005) describe, but do not endorse, a position that has the form of generative atomism.

not always invertible—the fact that a method of analysis is not always the simple inverse of a method of synthesis is well known to chemists and cooks. Finally, the analytic process may fail because the system has no atoms. That is, each domain of the ontology is constructed from elements of other domains that precede it in the construction hierarchy, but there is no lowest level to ground the process. One example of a system that allows decomposition but has no atoms is a Wada Basin fractal. This can be generated by placing four perfectly spherical mirrors in a tetrahedron and positioning a light source at the center of the space between the spheres, producing an infinite regress. Atomless Boolean algebras constitute another form of structure that violates one of the conditions for generative atomism.

Generative atomism can thus fail in the following ways. There can be nonfundamental entities that are not constructed from the atoms of the chosen domain. There can be nonfundamental entities that are not analyzable into atoms of the chosen domain. There can be no atoms in the chosen domain. And it can be impossible to formulate generating rules to construct all nonfundamental entities from the atoms of the chosen domain. Each of these types of failure except the third can result in the occurrence of a distinctive type of emergent entity.

Generative atomism is a formidable position, the enormous advantages of which are most directly evident in its ability to represent systems and their states by formal, extensional, mathematical techniques—something that is made possible by a parallel between the generative aspects of the systems and the generative aspects of the representations we use. Generative atomism leads to the in-principle predictability, the explainability, and the lack of novel features of the whole system with respect to the properties of its parts. It also leads to the reducibility of the whole system to its parts and hence to a lack of autonomy of the compound systems. It is in these senses that generative atomism is antithetical to emergence, and the failure or inapplicability of generative atomism is a necessary condition for synchronic ontological emergence. In contrast, we shall see that synchronic conceptual and synchronic inferential emergence, as well as diachronic ontological emergence, can be present within systems that possess the central features of generative atomism. Systems whose ontology satisfies the commitments of generative atomism are core examples of systems that are taken to lack the capacity for ontological emergence because the composite entities are nothing but structured groups of atoms. Thus, generative grammars, cellular automata, primitive recursive functions, mereological wholes and their parts, formalized proofs, mosaics, computer-generated embroidery, recursively axiomatizable theories, and many functionally characterized systems such as automobiles are familiar examples of generative atomism.

Generative atomism also lies at the heart of agent-based modeling in the social and ecological sciences. Those models are motivated in part by the tradition of methodological individualism in the social sciences—the view that all social phenomena are ultimately reducible to combinations of the actions of individual agents.[7] A common feature of all such systems is that everything about the compound entities is in some sense already present in the basic units of the system.

Generative atomism can be applied to a variety of types of entities: objects, properties, laws, functions, and propositions, among others, and a system can be amenable to a generative atomism approach with respect to one type of entity but not with respect to another. For example, a system can have an ontology of objects that satisfies the criteria for generative atomism but possess properties that do not.

Within those broad constraints, there are significant differences among examples of generative atomism. A well-known example satisfying both the analytic and the synthetic aspects of generative atomism is propositional logic, in which the syntax and semantics alike are generative; all molecular propositions are Boolean combinations of atomic propositions and the truth value of a molecular proposition is a function of the truth values of its atomic constituents. Conversely, every syntactically well-formed sentence can be uniquely decomposed into its constituent atomic sentences in a finite number of steps. Even this familiar and simple example of generative atomism involves an appeal to what can be done in principle, rather than in practice, because determining

7. In the case of social models, however, there is little reason to think that the relations involved in real social systems, as opposed to those used in the simplified models of the systems, are recursively specifiable.

whether a given Boolean sentence has a truth value assignment that makes it true (the satisfaction problem) has been shown to be an NP-complete problem (see Cook 1971). As a particular case, 3-SAT problems are NP-complete. That is, the problem of determining whether an arbitrary formula, in conjunctive normal form and in which each clause has at most three literals (an atomic sentence or its negation), has a truth value assignment that makes it true is in the class of nondeterministic polynomial (NP) problems (i.e., any proposed solution can be assessed for correctness in polynomial time) and any other NP problem can be transformed in polynomial time into the 3-SAT problem by suitable transformations of the inputs. NP problems have no known efficient solution method, at least in the worst-case instances, and so exact solutions are viewed as practically intractable. Such cases provide examples of in-practice unpredictability.

Atomism and, a fortiori, generative atomism are at best contingently true of nonmathematical subject matters.[8] So, their truth in any particular nonmathematical application must be a matter for empirical discovery rather than for metaphysical stipulation. This is a primary reason why a metaphysics of emergence informed by science, rather than by pure a priori analysis, is required. Conversely, to show the inadequacy of a universal commitment to generative atomism we must show that there are empirical features of the world that cannot be accommodated by that approach, or we must show that certain objective features such as explanatory power are missing when atomism is adopted.

1.3.1 An Aside on Eddington's Tables

Because of its combinatorial nature, generative atomism leaves little, if any, room for genuine novelty. Everything in the world is a result of rearrangements of the fundamental constituents with the only novelty being that of arrangements of those constituents. Solutions to some famous philosophical problems can be seen when we are alert to the limited scope of generative atomism. The classic problem of Eddington's two

8. In contrast, it is necessarily true of subjects such as Boolean algebra (on a Platonic interpretation).

tables (Eddington 1929), although usually posed in terms of realism, is directed at the tension between reduction and emergence. Eddington put the question this way: Which of the objects in front of us is the real table—the familiar, middle-sized object possessing rigidity, solidity, brownness, and woodenness, or the extraordinarily complex collection of atoms that make up the table? Although Eddington's challenge beautifully captures one of the core issues of reduction, it is clearer if we formulate the question in terms of properties rather than objects: Which of the properties of the object that we ordinarily call "the table" are real—rigidity, solidity, brownness, and woodenness, or the microscopic properties of charge, spin, and so on? It is not possible to answer that question until we have an answer to the basic reductionist question: Are the macroscopic properties reducible to the microscopic properties? One emergentist answer, and it is an answer that falls squarely into the realm of ontological emergence, is that, qua table, the emergent properties of rigidity and solidity exist in addition to the micro properties. These emergent properties are the result of long-range correlations between the atoms in the desk. These correlations produce global features, and it is simply wrong to think of the table as "nothing but" an aggregate of atoms. Whether other properties of the table, such as brownness and woodenness, are emergent is, as far as I know, currently an open question. The answer to Eddington's problem of the two tables is, thus, that it rests on a false dilemma generated by a tacit acceptance of generative atomistic physicalism (GAP). Both sets of properties—macroscopic and microscopic—are real and are needed for a complete account of the world.

In somewhat the same vein, Leibniz's famous mill example (Leibniz 1714/1989, section 17) gains its appeal from the instinctive tilt toward generative atomism. Leibniz likened the mind to a giant mill, and suggested a thought experiment in which a human was able to walk inside the mill. His suggestion was that nowhere in the mill would one be able to find something that explained perception. Mechanically and functionally, the mill is just another example of generative atomism, and the example works because there is nothing over and above the transparently combinatorial mechanism in the mill. Leibniz's example now seems quaint, but Searle's Chinese Room argument (Searle 1980) that the generative atomism apparatus of classical artificial intelligence is incapable of producing semantic understanding is a highly effective twentieth-century variant on Leibniz's earlier example. Although Searle's offhand remarks about semantic emergence are too cryptic to advance our understanding, his rejection of the generative atomism that is embedded in classical artificial intelligence is correct with respect to the property of semantic understanding.

1.3.2 Generalizations

Mereology is often conceived of as a core case of generative atomism, but despite its ambitions, it is not a general theory of the part/whole relation. David Lewis asserts that parthood is transitive: "a part of a part of something is always a part of it" (Lewis 1991, 3). But this is not true for functional parts. I am a functional part of the economy of Charlottesville, and my liver is a functional part of me, but my liver is not (yet) part of the Charlottesville economy. The counter-example might be avoided by insisting on the use of "X being a functional part of system Y, relative to function Z," but this would undermine the generality of the part/whole relation. For most people, and even for formalists, the last occurrence of "the" in this sentence is a grammatical part of the sentence, but its upper half is not, even though it is spatially part of the spatial part and of the spatial whole. It is this need to specify the type of parthood involved that undermines the ambitions of mereology to be a general theory. The failure of mereology for functional composition is one reason systems engineering has been forced to consider emergent features.

Moving to more general types of generative atomism, by GAP I mean an ontological position in which the fundamental entities are physical atoms in the sense described earlier, and structured arrangements of these atoms determine the entire ontology.[9] Two common choices for how the nonfundamental ontology is determined are the use of recursively specified rules of combination and some form of determination relation between the atomic structure and the remaining parts

9. I leave aside here treatments of "gunk"; see, e.g., Arntzenius (2008).

of the ontology. Well-known examples of such approaches are David Lewis's program of Humean supervenience (Lewis 1986a), the position that all properties are either basic or structural (see O'Connor and Wong 2005, although they do not accept GAP as a comprehensive position), the combinatorial ontology espoused by David Armstrong (Armstrong 1997), and I shall argue, the British Emergentist positions of C. D. Broad and Lloyd Morgan (1923).

Rather than a metaphysical position, GAP should be considered as a methodological approach—treat the ontology of the system as one covered by GAP until evidence suggests otherwise—that works well for certain systems but is inadequate for other purposes. As a statement about ontology with universal scope, it is false, as we shall subsequently see. As a first step to establishing this, I shall construct a system possessing features that violate the conditions for GAP. The purpose of this artificial system is to provide a clear and simple example from which the principal conceptual results can be drawn. From there, we can proceed to more complex real-world examples.

1.4 CHECKERS WORLD

Consider a very simple world, one that is very different from ours. It has a discrete finite spatial and temporal framework consisting of an 8 × 8-square grid, and its states are structurally identical to those of a single checkers game.[10] The elementary particles in this world are of two kinds, black and white, and their behavior is described by rules very similar to those that apply to checkers: in the early stages of this universe, the particles always move away diagonally from their initial position one square at a time; when diagonally adjacent in the forward direction to a piece of the other color, they move to the square immediately past the other piece on the diagonal and that other piece is annihilated, and so on. There is one important difference between this Checkers World and

10. By "checkers" I mean the standard American checkers game, also known as draughts in England. International draughts uses a 10 x 10-square board, but the game is otherwise similar.

checkers. The difference is that checkers is a game, the rules of which functionally define it, in the sense that you can use any kind of object as the pieces—donuts, precious gems, copies of books by Camus and Sartre—as long as they are of two different types and they retain their type identities from move to move. In contrast, in Checkers World, as a matter of nomological necessity, only black and white elementary particles are governed by these laws. The black and white particles are atoms of the system. They are immutable in the sense that the properties of each type of particle, as enshrined in the laws that apply to them, remain the same regardless of the positions of the other particles. Of course, some of those properties, such as the ability to annihilate a particle of the other color, only manifest themselves when other particles are suitably positioned, but any given black or white particle permanently possesses the disposition to annihilate particles of the other type. By stipulation, each type of particle is indivisible in that no proper parts of the particles can exist separately from the whole particle. Finally, the two kinds of particles are distinguishable from one another, both at the type and at the token level.

Now consider the entire temporal sequence of system states in Checkers World. The first N states consist in various spatial rearrangements of the elementary particles, with occasional annihilations. Then at time-step N + 1, a new type of particle, a blue particle or a red particle, appears when a white or a black particle reaches a square on, respectively, the boundary of the world opposite to where it started. Blue and red particles are not composed of two or more elementary particles, and they behave differently from how the black and white particles behave—they can move forwards or backwards and, as in a standard checkers game, blue particles can annihilate both black and red particles and red particles can annihilate both white and blue particles. The red and blue particles are also atoms according to our criteria for atoms in section 1.5. Both immutability and distinguishability are straightforward; red and blue particles are not composed of white or black particles and are therefore indivisible. For those who think in terms of levels, this means that there are no interactions that can, by an inverse analytic process, create white or black particles from either blue or red particles. If we go on to count the blue and red particles as emergent, this irreversibility is simply a basic fact about this example. Other cases of emergence can be reversible.

There are some features of this Checkers World that are directly relevant to considerations of emergence. From an inferential perspective, it is not possible to predict, even from a full knowledge of the fundamental laws governing the black and the white particles and their positions up to N, what the states of the world will be after N. That is, the laws governing the fundamental black and white particles determine that a black particle will transform into a red particle when it reaches the lower boundary, and that a white particle will transform into a blue particle when it reaches the upper boundary, but those laws are specific to the domain of the black and white particles. All of the original laws governing the black and white particles continue to operate when blue and red particles appear, but those laws do not specify how the red and blue particles behave. To gain that knowledge, experiment and observation are required. The blue and red particles do place constraints on the black and white particles that were not present before the emergence of the former, but the laws governing the former operate alongside the laws governing the latter and are consistent with them. This mimics some aspects of "downward causation," although since there are no levels, the term is misleading.

Once in place, the ontology of blue and red particles is autonomous from the ontology of black and white particles, and the laws governing the blue and the red particles are irreducible to the laws governing the black and white particles. The blue and red particles are novel and appear as a result of the dynamics of the black and white particles, but there is no holistic aspect. There is domain-to-domain causation because the blue and the red particles can annihilate, respectively, the black and white particles, but even if one introduced levels in terms of temporal precedence, there is no problem of "downward causation" from blue to black particles because the domain of the black and white particles is not causally closed. Also, in Checkers World, there is no question of causal influence of a whole on its parts, even though there is domain-to-domain causation. Furthermore, even if we insisted that to make the laws of the black and white domain complete we had to include inter-domain laws that specified how interactions between entities in the black and white and the blue and red domains proceeded, that would still leave the intra-domain laws of the blue and red domains unspecified. What happens when a blue particle reaches the lower edge of the board? Nothing about the laws that govern the interaction of blue/red and black/white particles tells us. In this case, it is simply false that the properties and laws of the domain consisting of the lowest level, the black and white domain, fixes everything about the Checkers World.

A further conclusion can be drawn from this example, which is that the laws of the lowest level can be causally complete during one time interval and incomplete during another. Before the first blue or red particle comes into existence, the laws governing the black and white particles are complete; after that point they are not. Such a situation can be accommodated by adopting two principles: first, that laws connect properties, and second, that there are no uninstantiated properties.[11] Under that view, the properties associated with a red particle do not exist until there is at least one red particle to instantiate those properties, and without the relevant properties there are no laws about red particles.

1.5 ATOMISM

Two conditions are individually necessary for something to be an atom of a generative system: the atoms must be *immutable* (and as a result they are indivisible) and they must be *individually distinguishable.*

1.5.1 Immutability

The *immutability principle* has a diachronic dimension and a synchronic dimension:

> Atoms must retain their essential properties across time, and when embedded in larger units, the atoms have the same essential properties they have in isolation.

11. Both of these positions were advocated by Armstrong (1978a). I am of the view that both of these positions are correct, but I am not advocating them here.

Thus, a brick does not change its core properties when mortared into a wall, spark plugs retain their basic functional features when screwed into an engine block, and sentences retain their truth value when part of larger complexes (at least within truth-functional semantics). We cannot insist that all properties of the atoms are invariant under such embeddings—the brick stands in different spatial relations to other bricks once in the wall, the spark plugs will become hotter when the engine is running, and so on. Immutability is usually a property only of atoms because the principle is not always satisfied by entities that are generated from the atoms using the rules of composition. Consider a system built from rigid spheres that are connected by springs. The atomic spheres are immutable with respect to dimensions, but even the simplest compound system composed of two spheres connected by a spring is not.

The importance of immutability is apparent in the case of compositional semantics. What is sometimes, although perhaps misleadingly, called Frege's Principle requires that the meaning of a semantic unit such as a sentence be a function of the meaning of its constituents, such as the individual words.[12] One important aspect of Frege's Principle is its role in context-independent semantics, where the meaning of an individual term does not vary across the sentences within which it occurs. A second aspect is the claim that this meaning is possessed by the term in isolation when not occurring as part of a sentence. Notoriously, many terms of natural languages fail to satisfy one or both of these aspects of Frege's Principle, and this failure produces considerable difficulties in formulating semantical rules for such languages. We catch a glimpse here of how semantic holism and, possibly, semantic emergence can provide a heuristic for understanding emergence more generally. This heuristic will be used in chapters 2 and 3 for illustrative purposes.

1.5.2 Indivisibility

One of the principal historical meanings of the term "atom" is indivisible entity, and indivisibility as well as immutability were required of

12. The attribution to Frege was first made in Carnap (1947, 121). For a discussion of the principle. see Pelletier (2000).

atoms by Leucippus and Democritus.[13] Indivisibility is always indivisibility with respect to some property, and parthood is always indivisibility with respect to some identified space. We can then formulate the *indivisibility principle*:

> A is an atom of a domain with respect to some property C, and space S just in case A has no proper part, with respect to S, that possesses C.

The space S is often physical space, but for some entities it could be semantic space, time, or some other type of space. The property C can be intrinsic, relational, or functional and in the most straightforward case, can simply be the condition of existence.[14] In a cellular automaton, individual cells are the smallest units that have a state; in formal languages, it is logically impossible for a subsentential unit of language to carry a truth value; no part of a basic functional unit in a car, such as a fragment of a nut, can perform the function of the original unit; and according to our current state of knowledge, it is nomologically impossible for a proper part of a quark to carry its own charge.[15] The principle allows that the atoms are divisible in spaces other than S and with respect to properties other than C, because as we noted earlier, a formal symbol such as an occurrence of "the" can be physically bisected, but neither half can carry out the syntactic function performed by an intact token "the."

To take another example, supposing that there is a fundamental unit of charge, then that unit can be divided within some region of the space of logical possibilities, but the resulting imaginary parts cannot play a role in the actual laws of physics. Indivisibility is a special case of immutability, and I have chosen to discuss it separately here only because of its historical importance. Despite the historical emphasis placed on the

13. The post-Rutherford use of the term "atom" in physics is a courtesy use.

14. Those who hold the view that existence is not a property should ignore this case.

15. What constitutes an atom is unavoidably imprecise within many material contexts—a well-used nut can be damaged and still perform its function—but fundamental entities of physics seem to have sharp identity conditions even when their distinguishability is problematical, as do the fundamental entities of a properly constructed logical or mathematical system.

indivisibility of atoms, it is their immutability that is crucial for the generative atomism project because immutability guarantees a stable foundation for the ontology. The immutability of the atoms also entails that any analysis of a generative atomistic system ends with the identification of the atoms and their essential properties.

1.5.3 Distinguishability

Regarding distinguishability, we have the *distinguishability principle*:

> At the type level each object type must be distinguishable from every other object type and at the token level, each token must be unambiguously distinguishable from other tokens of the same type.

To take a familiar example, each primitive type of sign in a properly constructed formal syntax must be distinguishable from every other primitive type, and each token must be distinguishable from every other token. Poor handwriting violates both type and token distinguishability. Spatial separation usually forms the basis for distinguishability in the case of physical entities, although the distinguishability criterion is violated by some physical entities such as fermions, the state functions of which overlap considerably.

The distinguishability addressed by the principle is ontological, not epistemological. For large enough N, an N-sided polygon is indistinguishable by human perception or visual imagination from an N + 1–sided polygon, but they are ontologically distinct, and as Descartes noted for chiliagons, they are also distinct in the understanding.[16]

1.5.4 What Counts as an Atom?

In most circumstances, atoms should not be identified with whatever current physics identifies as fundamental. Because of the reference to a

16. Descartes (1641, Meditation VI). See also Mill (1884, 82). A chiliagon is a 1000-sided polygon.

specific property C in our definition of indivisibility, an entity can be an atom of one generating system and not of another. For the purposes of many social sciences, the atoms are humans or some other social agents such as insects and not subunits such as cells or molecules because the latter two are incapable of possessing social roles. This classification need not be merely conventional. If a human loses the property of rationality, he or she is ordinarily reclassified out of the social order and is often deliberately excluded from society.[17]

This relativization to a type of system is of great importance in model-based approaches to emergence, for the model rather than nature determines what is an atom. Yet even within the realm of ontological emergence, it is a remarkable feature of the natural world, exploited by various sciences and subareas such as effective field theories, that stable domains of reality exist. Within those domains can be found atoms possessing the features of immutability, indivisibility, and distinguishability. In some domains, the criteria for deciding these properties are less clear-cut than they tend to be in physics, but the classifications are not arbitrary.

We saw in section 1.2 that there are prima facie examples of emergence in the everyday world. These examples cannot be taken at face value as cases of emergence, precisely because the generative atomism tradition has been so successful at producing explanations at a finer-grained level of treatment. It frequently happens that a careful theoretical decomposition of any given system reveals that it is reducible to a composition of elementary parts. Thus, in cases where emergence apparently occurs, as when epoxy resin and hardener, both of which are malleable, are mixed and the compound mixture becomes extraordinarily hard, the appearance of this apparently novel and emergent property can be explained in terms of the interactions of the constituent molecules of the two components. The success of this finer grained analysis takes priority over the ordinary appearances and undermines our surface understanding of emergence.

17. By "rationality" I mean a very general feature of what it is to be human rather than narrower and perhaps prejudicial meanings such as are attached to "economic rationality." The extent to which irrational individuals are to be treated differently from rational individuals is obviously a contentious matter and involves how to deal with deviants ranging from sports hooligans to psychotic serial murderers.

But this success of generative atomism and its cousins is methodological. The principles of generative atomism are not necessary a priori metaphysical truths, and having enjoyed their status as received wisdom on the basis of their past empirical successes, they must be rejected as universal methods if the scientific evidence shows that they are empirically or theoretically unsuccessful in a number of currently well-established domains.

1.6 CRITERIA FOR EMERGENCE

A traditional goal of conceptual analysis with respect to a concept C is to formulate a set of necessary and sufficient conditions for C to correctly apply. In the case of emergence, there are too many different uses of the term "emergence" that are entrenched across various fields for a single comprehensive definition to be possible at this point in time or, perhaps, ever. Moreover, as noted earlier, unlike a topic such as causation for which there exist clear, commonly agreed-upon core examples, most examples of emergence are too remote from ordinary experience for us to have formed some kind of consensus on exemplary cases. So, an essential part of our task is to identify in as clear a way as we can wherein lie the rough boundaries of our subject.

A good place to begin that task is to identify criteria that are commonly used to identify emergent phenomena and to formulate a theoretical taxonomy that captures all, or most, of the current approaches. There are four features that recur again and again in discussions of emergence. Emergent features result from something else, they possess a certain kind of novelty with respect to the features from which they develop, they are autonomous from the features from which they develop, and they exhibit a form of holism.[18]

That is a descriptive claim. A normative claim that will be maintained in this book is that the first two features are necessary in all cases of emergence, whereas the autonomy and holism conditions are present

18. In Humphreys (1997b) I discussed six criteria for emergence. Although each of those six is still applicable, the four presented here form a more compact set.

in some cases of emergence but neither is necessary. Whether all four conditions are sufficient for an entity to be emergent is a question that cannot be answered in any general way. The reason is this: I have already noted that there exists a wide variety of approaches to emergence and that many of these approaches are addressing different properties of phenomena. Imposing order on this variety is not easy because claims of emergence occur when a desirable default property has failed and the breakdown results in having to switch domains. It is the breakdown of predictability or computability that usually leads to the unpredictability approach to emergence; it is the breakdown of generative atomism that often lies behind ontological emergence; it is the failure of constructibility or representability that leads to conceptual emergence, and so on. Indeed, the famous conflict between emergence and reduction arises because reduction is taken as an ideal that sometimes seems to fail and to result in emergence. It is instructive to think through whether your own favorite account of emergence is itself a result of the failure of some desirable method.

I have not included unexplainability in my list of criteria. Although emergent entities have often been considered to have no explanation for their existence in terms of the entities from which they emerge, this concedes too much to the mystical tradition of emergence. As we shall see, one of the striking features of contemporary work in emergence is that it allows us to understand how and why emergent entities emerge.

These four features give us a way to understand how otherwise disparate accounts of emergence seem to be drawing on a core group of features; and how, as those core features are made precise, the accounts of emergence diverge to produce the multiplicity of approaches to emergence that we now have. Indeed, one can already see how such traditional markers of emergence as unpredictability (epistemic novelty) and irreducibility (conceptual novelty, ontological holism) exemplify these features.

1.6.1 The First Feature: Emergence Is Relational

The first of these features—the fact that emergent entities must result from something else—may seem trivial, but it is important and frequently overlooked. The emphasis on the autonomy of higher level synchronic features and the common focus in the emergentist literature on phenomena such as consciousness can reinforce the idea that emergence is an intrinsic feature of a phenomenon. That this is wrong is shown by the clear difference between, on the one hand, the non-emergentist status of psychophysical parallelism, within which the mental realm is separate from the physical realm and events in the two realms occur in a synchronized fashion, and on the other hand, the situation that would occur if mental phenomena are emergent from brain states when a sufficient level of complexity is reached and interactions between neural elements produce novel properties. The requirement that the emergent feature must result from other entities also entails that dualism per se does not count as an example of ontological emergence, even though there are ontologically distinct and irreducible domains involved. There must be at minimum an appropriate kind of precedence relation between the emergent phenomenon and earlier or more basic entities.

However, the English-language expressions "emerges from" or "results from" are not quite what we need. The essential feature is that attributions of emergence have a logical form that is relational rather than monadic, and we must always consider the origins of the emergent phenomenon. The proper locution is "X is emergent with respect to Y," and not the common "X is emergent."[19] This relational aspect explains why we count neither synchronically fundamental entities as examples of synchronic emergence nor the spontaneous appearance of something from nothing as an example of diachronic emergence, because neither emerges from anything. Similarly, if God introduced botanical objects on the third day of creation, that does not count as the emergence of plants, even though they would be autonomous from that act and would depend for their continued existence on the water that was introduced on the second day.

Within the realm of ontological emergence, the relation has some distinctive characteristics. The relation is irreflexive and is rarely

19. Although for brevity, and when the second relatum is obvious, I shall use the latter for convenience.

symmetric, although with diachronic forms of ontological emergence and a correct understanding of novelty, cyclical patterns of states can result in a type A entity emerging from a type B entity and a type B entity then emerging from a type A entity. Transitivity is harder to assess. Unlike causation, where indirect causation is a well-established concept, emergence is arguably only found in a direct form. For cases of ontological emergence, it is interactions or processes in a domain B that lead to an emergent feature C, and in cases in which interactions or processes in a third domain A produce emergent entities in domain B, those A interactions are not the things from which the feature C emerges. Intransitivity can occur with other kinds of emergence. For instance, B may be unpredictable from A because A lacks a concept crucial for predicting B; and for similar reasons, C is unpredictable from B, but it does not follow that C is unpredictable from A.

1.6.2 The Second and Third Features: Novelty and Autonomy

Perhaps the most characteristic feature of emergent entities is that they are novel. A convenient way of identifying novelty with respect to some domain is to say:

> *Definition* An entity E is novel with respect to a domain D just in case E is not included in the closure of D under the closure criteria C that are appropriate for D. Novelty can be taxonomic novelty in the sense that the entity is not included in a given conceptual classification.

Here are five kinds of closure conditions, which I use as illustrations without endorsing them, especially the explanatory criterion:

(i) An entity is considered to be novel with respect to a theoretical base if and only if it is impossible to deductively predict its existence using the theoretical apparatus of that base. "Entity" here refers indifferently to objects, properties, states, and similar categories. Here, the closure condition C is deducibility or computability with respect to a set of rules. The existing theoretical apparatus is incapable either in practice or in principle of allowing a derivation of a statement asserting the existence of the novel entity, so the statement asserting the entity's existence is not in the deductive closure of the axioms of the theory. This leaves the approach open to the objection that an incomplete or an undecidable theory will have sentences about the subject matter of the domain that cannot be deduced from the theory and that such sentences describe emergent facts.[20] Rather than an objection, this point can be interpreted as reflecting the theory-dependence of prediction-based accounts of emergence—a phenomenon is predictively emergent relative to one theory T yet not with respect to another theory T′—and the accompanying difficulty of capturing what it means for a theory to be a preferred predictive apparatus.

(ii) An entity is considered to be novel with respect to a given explanatory base if and only if it is impossible to explain why the entity has the properties that it does using only the resources of that base. The closure condition C is explanatory closure under some specified model of explanation. Under deductivist accounts of explanation, such as Hempel (1965), as well as unificationist accounts such as Friedman (1974) and Kitcher (1989) this criterion subsumes category (i) above. Under ontological causal accounts such as Salmon (1984) and Humphreys (1989), the novel entity would fall outside the causal closure of the domain consisting of the explanans events or variables. Under this view there is no requirement that the emergent entity must be unpredictable.

(iii) An entity is novel with respect to a base set of entities if and only if a specified kind of dependency does not exist between that entity and entities in the base set. Here, the closure condition C could be closure under a supervenience relation such as logical supervenience. Examples of this use can be found in McLaughlin (1997) and Kim (1998).

(iv) A direct conflict between emergence and physicalism can occur because of a physicalist's commitment to the causal closure of the physical. One common definition of the causal closure of the

20. Gu et al. (2009) and Gu and Alvarado (2011) have used undecidability as a criterion for emergence. For a critical analysis, see Humphreys (2015).

physical asserts that the physical realm is causally closed just in case every physical event is fixed by some prior physical events and laws. This definition is inadequate because it does not preclude the existence of deterministic chains or processes that leave and re-enter the physical domain and have nonphysical events as intermediaries. To prevent this, we could require that the realm of the physical be causally closed just in case no causal chain or process from one physical event to another contains an event that is not physical. This, too, is unsatisfactory because it allows that there are causal chains affecting the physical that are initiated by nonphysical events or that produce nonphysical events. An emergent, nonphysical, psychological event could cause a physical event that in turn causes only other physical events. So I use the following definition:

> *Definition* Two events are causally connected just in case one is a cause of the other. Then, a domain D is *causally closed* just in case anything causally connected to an element of D is itself an element of D.[21]

Then causal novelty with respect to a domain D occurs when the causal closure of D fails. This allows the existence of non-physical epiphenomena. If such things are considered unacceptable on a physicalist position, they are easily banned by imposing a requirement that there cannot be a causally inefficacious physical entity.[22]

21. Because the definition of physicalism is not a settled matter, it is worth exploring a further distinction. One variety of physicalism says that the only substance is material or physical substance. Because substance is a philosophical posit rather than a scientific concept, what counts as material or physical is determined by the consensus of philosophical arguments. This substantival tradition allows that there could be nonphysical properties possessed by the material substance, but it bans substance dualism à la Descartes or substance multiplicity. The position, focusing as it does on substances, is consistent with the existence of emergent objects or properties, but because there are no convincing arguments for the existence of substances, this position will not concern us further.
22. That constraint is weaker than Alexander's Dictum, which bans any causally inert entity and disallows at one fell swoop all abstract entities. The requirement used here disallows time as physical and perhaps spacetime points.

As noted, causal closure is often confused with causal completeness. The latter is a distinct concept from the former and is properly defined as:

> *Definition* A domain D is *causally complete* just in case every event in D has a causally sufficient antecedent that is in D.

This definition allows that causal chains can enter and leave a domain, rendering the domain causally open, even though every event in D can be traced back via a causal chain to some event that is also in D.

(v) A final interpretation of novelty is that different laws apply to emergent features than to the features from which they emerge. We then have that entities of type B are novel with respect to a domain D if and only if there is at least one law that applies to type B entities that does not apply to entities in D. This definition does not require that type B entities are composed of entities drawn from D, although it is compatible with that situation. Novel laws can appear both diachronically and synchronically.

Each of these five types of novelty has played an important historical role in discussions of emergence, although the novelty has to be of the right type because the domain with respect to which the entity is novel may be insufficiently comprehensive. The domain with respect to which the entity is novel is usually taken to have some special status, such as offering a complete unifying theory in the prediction and explanatory cases, or providing a physicalist base in the causal case. Novelty is usually required to be of a striking kind—not just any novelty will do—but since what counts as an acceptable kind of novelty is rarely, if ever, specified, it is difficult to capture that additional aspect with any generality.

What of autonomy, our third condition?[23] Autonomy does not entail novelty in that A can be autonomous from B even though A and B are of the same type. Nor does novelty entail autonomy. As one example, the effect of a cause can be novel with respect to that cause, but the later

23. Authors who have argued for autonomy as a criterion for emergence include Holland (1998, 7), Batterman (2002), and Abbott (2006).

event is causally dependent on the earlier and hence not autonomous. However, autonomy and novelty are frequently associated. Predictive novelty gives rise to predictive autonomy in the sense that a separate predictive apparatus is require to deal with the novel entity. The situation is similar with explanatory autonomy. In many cases of multiple realizability—cases in which, for example, a computational property can be implemented in a variety of different physical realizations—the conceptual apparatus of computability is not part of the conceptual apparatus of physics, chemistry, or biology. As a result, the computational level is explanatorily autonomous with respect to the domain of the physical. Causal and nomological autonomy can result in similar ways, but where a dependency relation holds between the emergent entity and its origins, there is less reason to hold that the former is autonomous from the latter with respect to properties that are determined by the dependency relation. If A logically or conceptually depends on B, A is not logically or conceptually autonomous from B. In contrast, when A is causally dependent upon B, A can be conceptually autonomous from B. It is because of this lack of autonomy when dependency relations are present that I have not required autonomy to be a necessary condition for emergence.

Returning to novelty, one interpretation of novelty to avoid is this: on a Platonist account of properties, there are no novel properties—only novel instances when the timelessly existing properties are first instantiated. Coupled with a theory of time that rejects genuine temporal change, there could be no ontologically emergent novelty in the sense of genuinely new properties or their instances. That view is encouraged by the kinds of logical and semantic convenience that are prevalent in generative atomism. This suggests that emergentists, and in particular diachronic emergentists, need an account of properties within which properties are features of the world rather than abstract entities and that novel properties are produced by interactions.

Moreover, we do not want to distinguish between the first and subsequent instances of a property with regard to their status as emergent. This "first instance" approach to novelty is often briefly canvassed and then rejected as being inadequate because it allows as emergent a mass of value precisely 1.000000045679345 kgm, a value that had never previously existed in the universe as a property of a single object.[24] We can accommodate this objection within our basic framework by insisting on a specification of what it is that the new value emerges from. If the new value of mass came about by a mere rearrangement of existing matter (by aggregating bits of smaller mass, or by dividing a larger mass), then by the conservation of mass principle, the "novel" property instance was there all along—it just was not instanced in the region that now has it.[25] When the new value was obtained by aggregation, it was a scattered collection of spatially separated regions that collectively had the supposedly new value of mass; in the case of division, it was part of the now larger region that had it. The novelty criterion for emergence would then be violated, and thus we have no case of emergence at all. Moreover, the "novel" value can be predicted from, explained by, and reduced to the masses of the parts by elementary applications of the procedures of generative atomism. This is an argument that can be applied to any combinatorial ontology in which aggregation and division takes place.

If we insist that a property never have had a previous instance in order to be emergent, that would then classify two qualitatively indistinguishable instances of the same property differently, the first being emergent and the second not. This is wrong—birth order matters for monarchs but not for emergents. Another way of capturing what it is for an entity to be different from others is this: an entity E is novel with respect to a domain of entities D just in case logically or nomologically E could not be a member of D.[26] The domain involved is not a set—membership in the domain is contingent, whereas a set necessarily has the members that it does.

As an example of the application of the definition, it is logically impossible for the property of being L-shaped to be possessed by the square floor tiles from which the L-shaped pattern is made. Here, E is the property "is L-shaped," D is the domain of instances of the property "is

24. For an early discussion of this point, see Teller (1992).

25. Assuming here that only mass, and not mass-energy, is involved.

26. This includes what Bedau (1997) has called the "nominal emergence" of a property, although Bedau does not specify the modal strength involved.

square," and E is novel with respect to D. It is nomologically impossible, and perhaps impossible in some stronger sense, for some of the properties of water to be possessed by hydrogen and oxygen atoms, and so those properties are novel with respect to the domain of physical atoms. In the molecular domain, the liquidity of water, a prime candidate for an emergent property, is something that cannot be possessed by individual molecules of H_2O. In the case of ferromagnetism, the novel property is the permanent possession above the critical temperature of an overall magnetization of macroscopic iron, which cannot be possessed by the individual dipoles in the metal. I am not claiming here that these properties are emergent, even though the second, third, and fourth examples are; only that they are novel with respect to a given domain. The definition also gives us this correct consequence: if we have three qualitatively indistinguishable instances A, B, C of a given entity, where A is emergent with respect to some domain D, and B results from A by a process of replication, whereas C results from D by means of the same kind of process that produced A, then C is emergent with respect to D, but B is emergent with respect neither to A nor to D.

1.6.3 Holism

Traditionally, holism—summed up in the slogan that "the whole is greater than the sum of its parts"—has been a core part of what is wanted from emergent phenomena.[27] Emergent phenomena are traditionally differentiated from what are called "aggregate phenomena." The idea is that, in contrast to emergent entities, aggregate entities are nothing but their constituent units, a particularly clear example being the building material called aggregate, which is simply a collection of small stones used in making concrete. Aggregated entities can be structured, but this structure is not permitted to lead to the development of a distinctively new kind of property. Aggregation can produce different values of a property already possessed by the constituents, and so the volume of

27. "[A]n emergent quality is roughly a quality which belongs to a complex as a whole and not to its parts" (Broad 1925, 23). The slogan dates back to Aristotle (*Metaphysics* 10f-1045a).

the building material has a value different from the volume of any of the stones in it, but this does not undermine the aggregative nature of the whole. Similarly, truth-functional compounds can have different truth values than (any or some of) their components, without the compounds losing their aggregate nature. This is reasonably clear, but not unimpeachably so, and it leaves the issue at the level of our intuitive grasp of examples. A huge heap of building aggregate has a type of property not possessed by any of the component stones—that of instability—and so aggregate can have non-aggregate properties. Indeed, as we saw in section 1.2.2, the instability of sand piles and related systems provides one of the core examples of emergent phenomena within complex systems theory. There is a cautionary tale here: emergent properties can appear in familiar medium-sized physical objects and not just in exotic quantum mechanical situations.[28]

Part of the unclarity of the slogan results from the widely acknowledged fact that reference to the "sum" of the parts of a system is both too restrictive—a system in which the parts interact according to a multiplicative principle does not, simply because of this, exhibit emergent features—and too vague. Still speaking casually, an emergent phenomenon could be less than the sum of the original parts, or there could be no proper parts in the whole.

So, let us retire this unfortunate slogan and take this definition as our starting point:

> *Definition* P is a nomologically holistic property of a system S just in case P is possessed by S as a whole, and as a matter of nomological but not metaphysical necessity, P could not be possessed by any of the proper parts of S.

Because holism tends to be a synchronic feature of an entity, this definition reflects the part/whole approach that is characteristic of many variants of synchronic emergence, but it can be applied to a diachronically emergent entity when that entity has proper parts. Holism is

28. An excellent case for the prevalence of various kinds of emergence in familiar phenomena is made in Wimsatt (2007).

insufficient for emergence because holistic properties can be possessed by systems that do not display emergent features. The number of objects in a finite collection of discrete entities is a property of the entire collection, rather than of the individuals that compose it, but this is no emergent property. More will be said about holism in section 3.5.

Are the red and blue particles of Checkers World, their properties, and the laws that govern them emergent according to these criteria? The red and blue particles develop from the behavior of and interactions between white and black particles, and they have novel properties in the sense that the closure conditions for the domain of white and black particles do not include the rules governing the behavior of the red and blue particles beyond their initial appearance. Those novel rules make the red and blue particles autonomous from the black and white particles. The blue and red particles are not composed of more basic entities, but such candidates for emergence as qualia and consciousness are not composite entities, either. Holism is thus not a feature of the red and blue particles. Given that three of our four criteria are satisfied, this is a reason to consider the red and blue particles as serious candidates for being emergent. Further arguments in favor of this conclusion will be provided later.

1.7 A TAXONOMY FOR EMERGENCE

Why is emergence important? The usual answer is that it results from a failure of reduction, and such a failure would require a significant shift in our ontological and methodological views. As I have suggested, one reason we have different accounts of emergence is that generative atomism, its inverse decompositional methods, and reduction more generally fail in a number of ways. A means to make this claim plausible is to give a descriptive classification of existing accounts of emergence and to show that a failure of certain aspects of GAP leads to the categories in our taxonomy. In that way, we have a philosophical explanation for a set of positions that would otherwise seem to have developed in a historically contingent fashion. I am not claiming that the development of these positions was consciously driven by an opposition to GAP, although

I shall argue that this was true for some contributors to the British Emergentist movement. The intellectual motivation for a position does influence the form that the position eventually takes, but my interest is in whether there is a coherent structure behind most of the accounts that have been considered emergentist. There is no guarantee that such a framework exists, but if we can find one, a better understanding of what it is for something to be emergent will be revealed. Our account is not stipulative; it is based on systematic accounts of emergence that are motivated by explicit arguments. It is appropriate to refuse to admit as a counter-example to the taxonomy some situation for which nothing has been offered as a justification but an intuitive belief that it ought to count as emergent.

A different approach, looking at how the word "emergence" is used and forming a classification on that basis, is not likely to be profitable. The term has casual, everyday uses—the man emerged from the cave—as well as precise uses, and it is often applied to phenomena without clear criteria for what counts as a correct use. In such situations, linguistic analysis is likely to end up with a heterogeneous mess, a semantic dog's dinner.

Our classification is two dimensional. Appealing to the criterion that one entity emerges from others, the *relational dimension* is based on relations between the emergent entity and the entities with respect to which it is emergent. There are three divisions in this dimension; ontological approaches, inferential approaches, and conceptual approaches. *Ontological* approaches consider emergent phenomena to be objective features of the world. Their emergent status, at least in cases that do not involve humans, is independent of our own existence and knowledge, and instances of ontological emergence occurred long before cognitive agents evolved in the universe. *Inferential* approaches consider the attribution of emergence to be a result of limited abilities to make predictions, not just by humans but also in a more objective sense. If a system is represented by a theory or a model R, and inferential emergence is present, knowledge about the emergent feature must be obtained from sources other than R—perhaps observation, perhaps experiment, perhaps via some other representational device. *Conceptual* approaches use novel theoretical and linguistic representations to represent features that are not efficiently describable in the vocabularies of other domains or that allow for more effective predictions. Although conceptual emergence is often linked to inferential emergence, the emphasis in the former is on representing an entity, whereas in the latter it is on derivability.

All three types tend to appeal, implicitly or explicitly, to a segmented view of the world that is divided into domains, with elements of one domain emerging from elements of other domains. This appeal to domains differs from the usual talk of "levels." The latter seems natural within an approach that has larger entities spatially including smaller entities as constituents, and this inclusion relation results in the familiar hierarchy of elementary particle physics, condensed matter physics, chemistry, biochemistry, biology, and so on. But this mereological conception of objects having smaller objects as components concedes too much to traditional positions based on generative atomism, and it is unsuited to diachronic accounts of emergence. It is thus better to use the more neutral approach of domains.[29]

The second dimension of emergence rests on the distinction between *diachronic emergence* and *synchronic emergence.* It recognizes that emergence can occur across time—what one might call the dynamic emergence of one kind of structure from another—as well as between two coexisting domains. This constitutes the *temporal dimension* of our taxonomy.

1.7.1 Inferential Emergence: Emergence as Lack of Derivability

I begin the relational dimension with inferential emergence. A representative version of this approach was given by C. D. Broad:

> The emergent theory asserts that there are certain wholes, composed (say) of constituents A, B, and C in a relation R to each other; that all wholes composed of constituents of the same kind as A, B, and C in relations of the same kind as R have certain characteristic properties; that A, B, and C are capable of occurring in

29. This suggestion to move from levels to domains is due to Jaegwon Kim.

> other kinds of complex where the relation is not the same kind as R; and that the characteristic properties of the whole R(A, B, C) cannot, even in theory, be deduced from the most complete knowledge of the properties of A, B, and C in isolation or in other wholes which are not of the form R(A, B, C). (Broad 1925, 61)

We have here features that are characteristic of an emergentist position—an appeal to a system that is composed of parts, yet which possesses a novel property that is underivable. In Broad's formulation, the "even in theory" condition indicates a potential ontological basis for the underivability, but underivability can also result from features of the theory describing the constituents. Holism is featured in this example, although it is not required in the inferential approach. Because it is not always easy to make clear sense of the idea of being underivable in principle even from an ideal theory, less stringent criteria using various measures of difficulty of derivation or computation are often used.

The inferential approach is the most common basis for accounts of emergence. It is the motivation for the weak emergence account of Bedau (1997), which was based on computational incompressibility, the mathematical singularity approach of Batterman (2002), the undecidability results of Gu et al. (2009), and the efficient prediction approaches of Shalizi and Crutchfield (2001) and of Gu and Alvarado (2011), some of which will be discussed in more detail in chapter 4.

1.7.2 Conceptual Emergence

Turning next to the conceptual approach, we see these accounts of emergence often result from a situation in which some form of inferential failure occurs. Conceptual emergence maintains that an entity, such as a state or a property, is conceptually emergent with respect to a theoretical framework F if and only if a conceptual or descriptive apparatus that is not a part of F must be developed in order to effectively represent that entity. This apparatus may include new predicates, a new law statement, or an entirely new theory; and the introduction of the new concepts will sometimes involve moving to a new scientific domain. The phrase "effectively represent" serves as a parameter for these approaches. For instance, the need to introduce the term "liquid" in order to effectively represent the collective behavior of H_2O molecules that have undergone a transition from the ice phase to the water phase is an example of conceptual emergence.

The primary relational aspect involved in conceptual emergence is one of conceptual irreducibility between the new and the old theoretical frameworks. Of our four criteria, the novelty of the concept will usually consist in being undefinable in terms of the concepts of F. Holism is often the underlying reason for needing the new vocabulary because the novel feature will frequently be of a kind that can only be possessed by ensembles of objects, rather than individual constituents, and these ensembles are sometimes structured, sometimes not. Although the novel concept does not emerge from other concepts, its use is usually required as a result of either the ontological or the inferential emergence of the phenomenon it describes with respect to a domain within which that concept is not applicable. Autonomy here is conceptual autonomy. Both the inferential approach and the conceptual approach apply to representations of the phenomena, rather than directly to the phenomena themselves. Examples of conceptual emergence can be found in Anderson (1972), Weinberg (1987), and Rohrlich (1997).

1.7.3 Ontological Emergence

Turning to the last part of the relational taxonomy, the ontological approach considers emergent entities to be genuinely novel features of the world, and their emergent status is independent of our knowledge. Ontological emergence can be subdivided into three kinds:

1. *Property emergence.* Traditionally this has been construed as asserting the existence of a hierarchy of levels such that at each level there is at least one property that is irreducible to combinations of lower level properties, and the higher level property is emergent with respect to those lower level properties. That approach is crucially dependent upon our being able to make objective sense of a hierarchy of levels, and we shall eventually develop a version of ontological emergence that is level free.

2. *Object emergence.* This makes a similar claim about concrete individuals and, derivatively, types of objects.
3. *Nomological emergence.* This asserts that there are distinctive and autonomous laws that govern specific domains that are distinct from the fundamental domain.

Candidates for ontologically emergent phenomena include entangled states in quantum mechanical systems, covalent bonding in chemistry, and ferromagnetism in condensed-matter physics. With ontological emergence, there is usually a clear sense that the new feature has emerged from other features, often as a result of interactions within a given domain, and that it is novel. Holism is frequently present, as is nomological or causal autonomy. The primary relata in the ontological approach are the emergent feature and the features from which it emerges, but the relation can be a more closely delineated relation, such as functional irreducibility (Kim 1999), ontologically interpreted supervenience relations (van Cleve 1990), or fusion (Humphreys 1997a).

The three approaches—ontological, inferential, and conceptual—are not mutually exclusive. For example, conceptual novelty is often forced upon us when predictability fails or is required when a new type of collective entity ontologically emerges. In turn, predictability can fail in cases where the states involve an ontologically new type of entity. Among the three of them, they cover every contemporary account of emergence of which I am aware, except for mystical views that consider emergent phenomena to be beyond the reach of understanding. I have not included in this taxonomy the position that emergent features are unexplainable, because a striking characteristic of most contemporary scientific candidates for emergent phenomena is that we *do* have credible explanations of why those candidates have their emergent features. They are thus essentially different from the examples of emergence cited in earlier eras, such as life, for which there was then no explanation, and likewise they are different from some contemporary candidates such as consciousness. No "natural piety" is needed to accept modern, scientifically based examples of emergence.

The fact that elements falling into each of these categories are relational is logically important, for rather than the criteria for emergence being applicable to an entity E in isolation, one must consider a domain D or a framework F with respect to which E emerges, in the sense that it is the relevant relations between E and D or F that determines the emergent status of E. The fact that some of these relations are negative relations, such as irreducibility, does not affect this general point.

1.7.4 The Temporal Taxonomy

The second dimension of the taxonomy, the temporal dimension, appeals to the differences between diachronic emergence and synchronic emergence, and it is easier to articulate. The tradition of synchronic emergence has been particularly influential in the realm of mental properties. It is commonly held that when a person has a specifically mental state in virtue of being in a particular type of brain state, the mental state is not reducible to the physical state, and that the former exists at the same time as the latter. One of the main themes of this book is that too much attention has been paid by philosophers to synchronic emergence at the expense of diachronic emergence.

Diachronic emergence primarily, but not exclusively, involves the emergence of novel phenomena from preceding phenomena as a result of a temporally extended process; synchronic emergence involves the simultaneous coexistence of novel "higher level" objects or properties with objects or properties existing at some "lower level," where the "levels" talk is best interpreted metaphorically. I note that it is possible to have both diachronic emergence and synchronic emergence occurring in a single process. This temporal dimension is orthogonal to the tripartite division of the relational dimension; each of those three divisions has diachronic and synchronic variants. For example, synchronic conceptual emergence can occur in theoretical contexts within which, according to a familiar tradition, Nagel's inhomogeneous reducibility (E. Nagel 1961, chap. 11) cannot be carried out. In contrast, when a material makes the phase transition from paramagnetism to ferromagnetism, the brief but temporally extended relaxation into the global ferromagnetic state involves a diachronic process, the end-state of which requires the introduction of a new concept, the long-range order, to fully describe.

Table 1.1 Examples of the six types of emergence.

	Inferential	*Conceptual*	*Ontological*
Synchronic	Broad 1925	Anderson 1972	van Cleve 1990
Diachronic	Bedau 1997	Rueger 2000a	Humphreys 1997a

A seminal example of each dual category is given in table 1.1, again with the caution that some of these positions occupy more than one category. (For example, both Bedau's and Broad's accounts of inferential emergence contain elements of conceptual emergence.)

Finally, we might canvass the view that we should abandon the quest to understand emergence in general and replace the concept with more specific features such as self-organizing systems, computational incompressibility, and so on. That is a defensible view, but to pursue it would be to overlook important connections between these other concepts. The term "emergence" does cover a variety of positions, but I hope to have shown that there is significant unity among those positions. I conclude by noting that this is a taxonomy of existing accounts of emergence. It is not meant to suggest that all of these approaches are satisfactory.

1.8 EXAMPLES OF EMERGENCE

One way to lessen the puzzlement people feel when faced with emergent phenomena is to show that emergent phenomena are more common than is popularly supposed and can, with some effort, be understood. It is possible that in pursuing this strategy, the reader will conclude "Well, if *that's* what you mean by emergentism, it's not surprising that things are emerging all over the place!"[30] This is a legitimate point to make, and that is why it is important to show that the accounts and examples

30. One might equally well respond to many existing accounts of emergence. "Well, if *that's* what you mean by emergentism, it's not surprising there's not much of it about, or that we can't understand what it is!"

of emergentism endorsed in this book conform to core ideas of emergentism that have been developed in the philosophical and scientific literature.

There are two separate questions we need to address. The first question: Is there a consistent, intelligible account of emergence that captures all, or almost all, commonly cited examples of emergence? The answer to this is no, as our later chapters will illustrate. The second question: Are there any plausible real-world examples satisfying at least one of these varied accounts? By a "real-world example" we need not restrict ourselves to naturally occurring examples. Artifacts and computationally generated examples are as real as cases that exist without human intervention. By "plausible" I mean examples for which we have well-confirmed theories or models that do not involve unmotivated idealizations. This condition is automatically satisfied for computational examples where the emergent phenomenon is a consequence of the computational model, but it needs additional evidence in its favor if the models are claimed to be about noncomputational systems.

At this early stage of the book, all of the examples described will be cases of diachronic emergence, and each should be taken only as a prima facie example of emergence. Whether they are genuine examples of the various types of emergence in our taxonomy will have to await the detailed treatment of those types in the following chapters. As a preliminary assignment, the type of emergence is given in parentheses: O for "ontological," I for "inferential," and C for "conceptual." The list of examples should convey the variety and pervasiveness of emergence.

1. *The formation and preservation over time of bird flocks.* Here, the flock, which is an object consisting of birds moving in a coordinated fashion in a pattern that maintains a distinctive if imprecise shape, results from the individual motions of the birds according to three simple rules of separation, alignment, and cohesion.[31] In terms of our four characteristics, the flock grouping results from the interactions between individual birds, its cohesion is a novel feature of the aggregate collection of birds, and that cohesion is a holistic aspect of the flock that cannot be

31. For details, see http://www.red3d.com/cwr/boids/.

possessed by individual members. The flock is also an autonomous spatial object in the sense that it is irrelevant which particular birds enter and leave the flock; new birds can replace the original members of the flock and the spatial pattern will persist. This is an example of diachronic inferential emergence in that the temporal evolution of models of flocking is computationally irreducible—before the model is known to be successful, there is no shortcut to predicting the global state of the flock, which includes the direction of its center and its spatial extent, other than running a step-by-step simulation. Conceptual innovation is also required through the use of the concept of "flocking."[32] There is as yet no clear evidence that real birds use such rules when they fly in groups. Similar models can also account for the schooling behavior of fish. (I, C)

2. *The appearance and dissolution of traffic jams.* This is often cited as an example of an emergent phenomena. Dynamically, the example is similar to the flocking example. Here, the clumping of cars emerges from the individual motions of the cars and a few simple rules about interactions between them—rules of minimal separation, the need to maintain a minimum differential of speed between adjacent cars, and so on. The clumps of cars form identifiable autonomous objects that are novel and holistic in the sense that individual cars cannot possess the property of being clumped. As with bird flocking, this is an example of diachronic inferential emergence. Once again, the emergent phenomenon occurs at the end of a computationally irreducible process—the only way to produce and predict the positions of the clumps is by running through the simulation calculations. "Predict the positions of the clumps" means (i) that the exact position and time of formation of the clumps in a specific traffic flow cannot be predicted and (ii) that before enough examples of traffic jams in specifiable types of circumstances, such as the presence of an entry ramp onto the road, have been observed, the very existence of such clumps could not have been predicted. In this example there is evidence that some of the agent-based models of traffic jams are realistic models of actual traffic behavior (Sugiyami et al.

32. Perhaps a case can be made that some form of synchronic emergence results from the maintenance at the group level of features that result from the multiple complex interactions that occur at each stage, but I shall not defend that here.

2008). Models of traffic jams that use hydrodynamical theories have different properties and can provide compressible predictions. (I, C)

3. *First-order phase transitions*. Examples of these are the transition from a liquid to a solid crystal and the transition from water vapor to liquid water. These examples involve small idealizations in that, for example, sharp discontinuities only appear in pure water, with impurities being responsible for slush in everyday samples of water. Here, the emergent properties in the two cases are, respectively, rigidity and liquidity. In these examples, a quantity called the "order parameter," which captures the degree of organization in the system and is in this case the long-range order, undergoes a discontinuous change at the transition point. A transition in either direction, from water to steam or from steam to water, constitutes a phase transition. There are significant questions about whether a transition from an ordered state to a disordered state should count as a case of emergence, not the least because the disordered state results from a lack of interactions between the components of the system. Imposing such a constraint would rule out the transition from water to steam as a transition involving emergence, even if it is a phase transition. In addition, theoretical models of many phase transitions require an essential appeal to infinite limits, whereas the physical systems they model are finite and the appropriateness of this is a source of considerable dispute (Batterman 2002; Butterfield 2011a, 2011b), but the models are generally regarded as among the best understood of the candidates for emergence. The four features of the ordered state emerging from the disordered state, the autonomy of the ordered state from the disordered state, the novelty of the property on the other side of the phase transition relative to the original property, and the holistic nature of that property are all present. (I, O)

4. *The Belousov-Zhabotinsky (B-Z) reaction*. The original example of this chemical reaction occurred when citric acid was placed in a sulfuric acid solution in the presence of a cerium catalyst. This gives an oscillatory color change from yellow to colorless to yellow, and takes about an hour to reach equilibrium. These patterns dynamically emerge from the initially homogeneous color; they are autonomous in the sense that the oscillations persist without external interference, they are clearly novel,

and they are a feature of the entire system.[33] The Brusselator model or the Oregonator model can be used to simulate this kind of oscillating chemical reaction.[34] (O)

5. *Bénard instabilities.* In these, a layer of fluid between two horizontal plates is heated from below and convection rolls begin. There is a critical temperature region within which the Bénard cells form. For smallplate separations, the dynamics are described by a strange attractor, and there is a reduction in the degrees of freedom that leads to the emergence of rolls (Bishop 2008). Because nonlinear systems often have multiple attractors, there will often be more than one final state for such a system. To take two such examples, there can be two different possible directions of magnetization in a ferromagnet or two different directions of circulation in a Bénard roll. Which of these end states occurs is dependent upon chance fluctuations near the beginning of the self-organization process, and so it is unpredictable in which state a system will finish. We thus have a connection with the inferential mode of emergence. (O, I, C)

Couette flow is a similar example of a dissipative system far from equilibrium and occurs when a fluid is placed between two rotating cylinders having different rotational velocities. When the velocity gradient exceeds a critical value, rolls of vortices form. In each case, we have the features of structure emerging from homogeneity, novel features that are holistic and autonomous both in the sense that which fluid molecules are involved is irrelevant and in the sense of being realizable in different types of fluids. (O, I, C)

6. *Stigmergy.* This is a method of communication in emergent systems in which the individual parts of the system communicate with one another by modifying their local environment. A common example is the use of pheromones by ants to mark their trails and by termites to build their nests. In the latter case, termites, attracted by the pheromones,

33. Another example of a B-Z reaction is malonic acid oxidized in an acidic medium by bromate ions with cerous or ferrous ions as a catalyst. It also gives rise to spiral waves of oxidation in the unstirred layers of the B-Z reagent. The latter are similar to waves of excitation in neural or cardiac tissue (see Strogatz 1994, 255).

34. The equations for the Brusselator model are: $dx/dt = 1 - (b + 1)x + ax^2$, $dy/dt = bx - ax^2y$ where $a, b > 0$ and x, y are dimensionless concentrations.

tend to preferentially make mud deposits at sites where other ants have already made their deposits. Stigmergy is analogous to niche construction, in which species modify their environment and thus alter the evolutionary context. Termite mounds result from the interactions of pheromones and individual ants, they are novel types of structures that are a holistic consequence of colony activity, and they are autonomous in the sense that which individual ants built them is irrelevant. (O)

7. *Schelling's model of segregation* (Schelling 1978).[35] This famous model involves two types of individual, each type having a small but not overwhelming preference to have a majority of neighbors like themselves. With a random initial distribution of the two types on a two-dimensional grid, the types dynamically self-segregate as a result of moves to locations where their preferences are satisfied. The emergent feature is the novel qualitative property of spatial segregation of the two groups, which results from self-organization starting with an initially random spatial distribution. The property is autonomous in the sense that it is independent of which individuals are involved, and is self-sustaining under replacement of those individuals. The spatial property is essentially holistic and is not found at the individual level. (I, C)

Related examples are the oscillations of firefly flashes, Adam Smith's "invisible hand" of economics through which optimal pricing in markets results from many local interactions among economic agents, the spread of fads in a society, and the ability of self-organizing slime molds to form a spore-bearing structure.

We shall see more examples of emergence throughout the book, but these examples are enough to provide substance to our previously somewhat abstract discussion. And they reveal two things: first, that emergent properties are quite common in the physical and social realms; and second, that the existence of these detailed models at the physical level dissolves the air of mystery that has traditionally surrounded emergentism and has led clear-headed philosophers to stay away.

35. Research by Rainer Hegselmann has revealed that a very similar model to Schelling's was earlier developed by J. M. Sakoda. For details of a fascinating episode in the dynamics of scientific creativity, see Hegselmann (forthcoming).

1.9 OTHER APPROACHES TO EMERGENCE

There are other taxonomies that could be imposed on the emergentist literature. A common distinction made in the emergence literature is between weak and strong emergence. Although the use is not uniform, a characteristic feature of what is referred to as strong emergence is that there exist novel, irreducible, causal powers possessed by the emergent feature that result in downward causation.[36] To make this description slightly more precise, an entity, such as a property or an event, is strongly emergent only if it has the ability to causally influence entities at a lower level by means of features that belong to the emergent entity but that are not found in structured aggregates of lower level entities. In the case where the affected level is the physical level, the existence of downward causation is said to violate the causal closure of the physical domain. For this and other reasons, it is widely held that strong emergence is either an incoherent or an unscientific position.[37] I have avoided this classification into weak and strong emergence for two reasons. The first is that it commits us to the levels picture, a metaphor that is best avoided when possible on the basis that precise criteria for identifying levels are difficult to construct and that the metaphor has led to an undue emphasis on synchronic accounts of emergence. The second reason is that there is no need for a defender of ontological emergence to commit himself to downward causation and the problems that follow, as we shall see with fusion and transformational emergence.

Silberstein and McGeever (1999) have used the category of epistemological emergence. Examples that fall into the category of

36. "Strong emergentists maintain that evolution in the cosmos produces new, ontologically distinct levels, which are characterized by their own distinct laws or regularities and causal forces. By contrast, weak emergentists insist that, as new patterns emerge, the fundamental causal processes remain those of physics" (Clayton 2004, 9); also, "F is a (weakly) emergent property of S if (a) there is a law to the effect that all systems with this microstructure have F; but (b) F cannot, even in theory, be deduced from the most complete knowledge of the basic properties of the components $C_1, \ldots, C_N$ of the system" (10).

37. "All the evidence today suggests that strong emergence is scientifically irrelevant. . . . There is no evidence that strong emergence plays any role in contemporary science. . . . Strong emergence starts where scientific explanation ends" (Bedau 2002, 11). See also Ladyman et al. (2007) and Kim (1999).

epistemological emergence contain elements of both inferential and conceptual emergence, as can be seen from their definition:

> A property of an object or system is epistemologically emergent if the property is reducible to or determined by the intrinsic properties of the ultimate constituents of the object or system, while at the same time it is very difficult for us to explain, predict or derive the property on the basis of the ultimate constituents. Epistemologically emergent properties are novel only at a level of description. (1999, 186)

Epistemological emergence is a valid category, but it is preferable to keep separate the inferential and conceptual categories because there are pure examples of each.

1.9.1 Emergence as Unexplainability

This approach asserts that a property P is emergent if it is impossible to explain the occurrence of instances of P on the basis of any ideal theory.[38] What counts as an ideal theory is often left unarticulated. Under accounts of explanation, such as Carl Hempel's, that subscribe to the symmetry of explanation and prediction, the view that every adequate explanation could have, before the fact, served as a prediction of the outcome, and vice versa, emergence as unpredictability and emergence as unexplainability are equivalent. On that view, the unexplainability approach is a version of an inferential account. The symmetry thesis is no longer widely held, and this allows the possibility of explaining an emergent phenomenon even when it is theoretically unpredictable. In particular, the complexity of interactions between agents may not allow us to predict the state of social systems far into the future, but once we know what that state is, we can in some cases understand why it appeared, post hoc, in terms of the

38. For example: "Crudely speaking, a property P of an object O is emergent if P is the result of physical properties of the physical constituents of O while at the same time it is impossible to explain P in terms of the properties of the physical constituents of O" (Newman 1996, 245–46). Samuel Alexander also claimed that emergence "admits no explanation" (Alexander 1920, 2:47).

rules governing the system and the initial state.[39] Depending upon one's account of explanation, the impossibility of an explanation could either be due to a lack of ontological grounds for the explanation, such as the unavailability of a causal basis for explanation, or be based on conceptual difficulties, in the sense that conceptual innovations are required to understand the novel phenomena.

1.9.2 Nomological Emergence

This approach emphasizes the role of novel scientific laws in emergent phenomena. It asserts that entities of type B are emergent from entities of type A if and only if entities of type B have type A entities as constituents and there is at least one law that applies to type B entities that does not apply to type A entities. This definition requires a commitment to a synchronic compositional ontology, and I shall provide an alternative diachronic approach in chapter 2.

The classification of this approach depends upon one's view of the status of laws. If laws are objective features of the world, then novel laws will be examples of ontological emergence. If new law statements are a result of the introduction of new conceptual resources, then we have conceptual emergence. The existence of law statements that were underivable from the fundamental laws of a domain, but were applicable to entities within that domain, would suggest a case of inferential emergence.

1.9.3 Emergence as a Result of Essential Interactions

A common feature of an ontologically emergent property of an object O is that it results from essential interactions between the properties of the constituents of O—interactions that are nomologically necessary

39. For examples of causal theories of explanation that reject the symmetry thesis, see Salmon (1984) and Humphreys (1989). For an insightful discussion of the conditions under which facts about financial markets, conceived as complex systems, can be explained, see Kuhlmann (2014).

for the existence of the emergent property. A standard example of this type of emergence is ferromagnetism, which is discussed in detail in section 7.1. One motivation for insisting on a role for interactions is that we do not want what are called mere Cambridge changes to produce emergent properties.[40] So, the circus midget leaving the room results in my being the shortest person in the room, but I have not thereby acquired an emergent property, and the sense is that this is because nothing about me has changed. The kind of novelty relevant to emergence is, in contrast, a result of either spontaneous internal changes or interactions with others. Because interactions occur within systems that do not exhibit emergent features, as well as those that do, I shall not use this approach as either a criterion for emergence or as a distinctive kind of emergence, although we shall see that interactions play a central role in many cases of diachronic emergence.

1.9.4 Emergence as Undefinability

Teller (1992, 140–41) has suggested that a property is emergent if and only if it is not explicitly definable in terms of the non-relational properties of any of the object's proper parts. This undefinability approach is motivated by anti-reductionism, and falls squarely in the realm of conceptual emergence because the undefinability of P entails that A is conceptually novel. It suggests that if something—say, density—is explicitly definable in terms of (what we can suppose to be the intrinsic properties of) mass and volume, then because explicitly defined predicates are eliminable from the theoretical language, the corresponding properties need not be considered as real.

If we take the appeal to definitions to involve nominal definitions, then this criterion is linguistically based and classifies predicates as emergent rather than emergent properties directly. Whether or not

40. A Cambridge change occurs with respect to an entity *e* just in case there are times t_1 and t_2 such that t_1 is earlier than t_2 and some predicate P correctly describes *e* at t_1 and does not correctly describe *e* at t_2. A mere Cambridge change with respect to *e* is a Cambridge change with respect to *e* that is not a real change in *e*. What constitutes a real change in *e* requires more analysis than is required here; for present purposes we can take it as a change in at least one intrinsic property of *e*.

a predicate is definable in a given language depends upon the logical resources of the language used. For example, the identity relation, the property of "having measure zero" and the property "is a random variable" are not definable in first-order logic, but they are definable in second-order logic. This linguistic relativity makes an application to ontological emergence difficult.

1.10 THE RARITY HEURISTIC

A tradition in the emergentist literature has held that ontologically emergent phenomena are rare, consisting in things like life and human consciousness. This tradition can be summarized in a widely held heuristic:

> *The Rarity Heuristic* An account of emergence that makes ontological emergence commonplace has misidentified the criteria for emergence.

One of the themes of this work is that the Rarity Heuristic is incorrect, and that ontological emergence, while currently plausible only in a number of physical and chemical examples, is not uncommon. One argument that is used to defend the Rarity Heuristic is the remarkable success of generative atomism, as well as analytic methods more generally. Its successes in many areas of the natural sciences are well documented. On a more practical level, mass manufacturing and the assembly line would not have been possible if compositionality did not work well in the everyday world. For instance, a car really is composed of a few thousand parts arranged in a functionally appropriate sequence of operations and can be readily decomposed into its constituents.[41]

As appealing as the Rarity Heuristic is, insisting on its truth hinders us from seeing how emergence works. Rather than there being a gulf dividing emergent features from their bases, emergence can come in small increments, and this calls into question the truth of the Rarity

41. As long as chemical reactions and electromagnetic processes involving the fuel, the battery, and a few other components are not included.

Heuristic. The reader may find unsatisfying the notion that emergence occurs often, and that there are many more levels of phenomena than are traditionally conceived. Yet it is hard, given the evidence, to justify the Rarity Heuristic as anything but a relic of the failed attempt to force-fit all systems into the generative atomistic approach; and it is plausible that we shall eventually understand various phenomena in terms of progressively more complex interactions between progressively more complex entities. The use of effective field theories (EFTs), for example, is consistent with the idea of a nested series of models within high-energy physics.[42]

The Rarity Heuristic should thus be employed with caution and not used as normative principle to reject particular accounts of emergence.

42. See Castellani (2002) and Hartmann (2001).

2

Ontological Emergence

2.1 PRELIMINARIES

Ontological emergence asserts that some emergent features or objects are real features of the world and their status as emergent does not depend either on the state of knowledge of cognitive agents or on the degree of sophistication of the representational apparatus employed by such agents. This independence from cognitive agents allows ontologically emergent features to be objective and autonomous components of the world, rather than artifacts of a representational apparatus or of our epistemological limitations.

Ontological emergence is the most controversial form of emergence, and many writers deny that it can consistently be defended, especially when it involves downward causation. Although ontological emergence is often identified with strong emergence (see section 1.9), for which the identifying feature is the presence of downward causation, this identification is incorrect. There are ontologically emergent properties within physics, which is often considered by philosophers to constitute a single domain for ontological purposes; and these ontologically emergent properties can cause effects that lie entirely within the emergent property's own subdomain of physics. In section 2.2, I shall describe one way in which this can happen. When downward causation does occur, it need not produce difficulties and one way in which this can happen will be demonstrated when fusion emergence is introduced in section 2.3. It is likely that there are other kinds of ontological emergence beyond those I describe here, but providing a coherent account of one set of related approaches should reduce, if not eliminate, the sense of unease often associated with ontological emergence.

In this chapter I shall develop and defend two accounts of diachronic ontological emergence, transformational emergence, and fusion emergence, and provide what I believe to be convincing examples of each. Before doing so, the underlying ideas can be conveyed through a plausible but provisional example.

2.1.1 A Prima Facie but Probably Not Genuine Example of Transformational Emergence

It will be useful to begin with an everyday example to convey the general idea behind transformational emergence, and then proceed to a more detailed characterization. I have chosen the example of a mob because it strikes a balance between being intuitively accessible and being open to a reasonable degree of analysis. It also has the advantage that a mob has often been cited as a plausible candidate for exhibiting emergent features. Whether the emergence is real rather than apparent requires a considerably more detailed knowledge of social and individual psychology than we currently possess. So, the example should be viewed as a guide to understanding the concept of transformational emergence, a particular kind of ontological emergence that is the centerpiece of the next section, rather than as a definitive example on either side of the debate about the existence of ontological emergence in the world.

A mob is a loosely structured group of humans manifesting antisocial behavior. There is an important difference between mobs and crowds. Although both exhibit a spatial cohesiveness similar to that possessed by self-organizing flocks of birds—a case that is often cited as an example of some form of non-ontological emergence—and in all three cases (mobs, crowds, and flocks) there are relations between the members that are responsible for this spatial cohesion, it is not this spatial feature that is the core property of mobs. Mobs exhibit distinctively group-level properties—properties peculiar to social psychology that include the manifestation of extreme levels of violence. These properties have often been cited as examples of irreducible, emergent properties; as a result, mobs were the basis for early discussions of methodological individualism. These discussions revolved around whether mobs have group properties that are not reducible to the properties of individual members of the mob.

Here is one characterization of methodological individualism by John Watkins:

> [M]ethodological individualism is contrasted with sociological holism or organicism. On this latter view, social systems constitute "wholes" at least in the sense that some of their large-scale behavior is governed by macro-laws which are essentially *sociological* in the sense that they are *sui generis* and not to be explained as mere regularities or tendencies resulting from the behavior of interacting individuals. (Watkins 1957, 106)[1]

We see in this quote from Watkins a move from the individual to the group and to the associated idea that emergence, if it exists, results from holistic features of the group. This emphasis on the holistic properties of mobs and the role played by irreducible sociological laws is understandable, but secondary. The origins of at least one important property of mobs lie instead in the changes that many of its members undergo as a result of interactions between those members at the individual level. The salient feature of mobs is that individual members of the mob change from being psychologically normal human beings before the mob forms to psychologically abnormal individuals once they have interacted with other members of the mob. It is these interactions and the transformations which result from them that are the crucial feature of transformational emergence.

When humans transition to operating within mobs, they temporarily lose some of the features that make them civilized humans—they shift to primitive modes of behavior with diminished levels of rationality and a tendency to brutal behavior. They retain their identity as physical and biological entities, but they have lost their identities as

1. Compare this with Ernest Nagel's characterization of methodological individualism: "the methodological assumption that all collective terms designate either groups of human individuals or patterns of human behavior leads to a more fruitful way of identifying the extensions of such terms than does the perplexing hypostasis of mysterious super-individuals" (E. Nagel 1961, 537).

rational humans in virtue of interacting with other members of the mob, either individually, in small groups, or collectively. This shifts the individuals to a different domain within psychology, one for which the laws of the original domain have little applicability. You do not try to predict the behavior of members of an enraged mob by using rational choice theory.

Contrast the case of mobs with the case of flocking birds. Although the birds do behave differently within the flock as a result of interactions with other birds, nothing essential changes about the birds. This is true of many aggregates of humans, as can be seen from the example of smart crowds—groups that form as the result of electronic communications between the individual members, but in which no radical transformation of the members takes place.[2] In such latter cases, we have a good grip on at least a part of the processes of self-organization, and although the mechanisms that lead to a smart crowd initially forming are different from the mechanisms that lead to interactions once the individuals are a part of the crowd, we can for simplicity assume that the formation mechanisms are also responsible for the continued cohesion of the group. This self-organization may lead to inferential emergence or to conceptual emergence, but ontologically, we can completely analyze the system in terms of the original constituent agents and their interactions. The individuals in a mob have, in contrast, changed to a different type. With this in mind, we can now formulate the central ideas of transformational emergence.

> A man who drinks too much on occasion is still the same man as he was sober. An alcoholic, a real alcoholic, is not the same man at all. You can't predict anything about him for sure except that he will be someone you never met before.
>
> Raymond Chandler, *The Long Goodbye*

2. The term "smart mob" is more commonly used, but it is inaccurate because most examples of this genre are peaceful. I am assuming here that the communications that organize the crowd are not transmitted from a central source but, rather, are spread spontaneously from individual to individual.

2.2 TRANSFORMATIONAL EMERGENCE

I shall formulate transformational emergence in terms of individuals because that is the easiest to grasp, but the general approach can be applied to states, properties, and perhaps other kinds of entities.[3] Here is the basic idea: Transformational emergence occurs when an individual *a* that is considered to be a fundamental element of a domain D transforms into a different kind of individual a^*, often but not always as a result of interactions with other elements of D, and thereby becomes a member of a different domain D*. Members of D* are of a different type from members of D. They possess at least one novel property and are subject to different laws that apply to members of D* but not to members of D. The fact that a fundamental entity can be transformed makes transformational emergence essentially different from generative atomism, in which the fundamental entities are immutable.

Because of these transformations, collections of what were originally *a* type entities can appear to have holistic properties when the *a* type entities are mistakenly construed as retaining their former identities, although those group properties are actually aggregate properties of the transformed type a^* individuals. In particular, the sociological laws of mobs mentioned by Watkins are not irreducibly sociological; they are a consequence of the laws of abnormal psychology for individuals. In such situations, we are faced with two possibilities. If, by considering the characteristic laws and properties of a third domain $D^\dagger$, where $D^\dagger$ is disjoint from D and D*, it is possible to account for the novel properties of the transformed a^* individuals and the laws that they follow, in addition to accounting for the original properties and laws of *a* individuals, by using some form of atomism within the domain $D^\dagger$, then the transformed individuals a^* are not ontologically emergent with respect to D because the properties of members of D and of D* are both consequences of the properties and laws of $D^\dagger$.[4] But if, as a second possibility,

3. Elements of transformational emergence can be found in Humphreys (2011, 2014a). A related approach developed independently by Alexandre Guay and Olivier Sartenaer, and applied to the fractional quantum Hall effect, can be found in Guay and Sartenaer (2016).

4. The disjointness condition is to block the usual maneuvers of using arbitrary logical combinations of the original domains.

no such $D^{\dagger}$ exists, the a^* remain as primitive elements of D^*, and the a^* are novel with respect to D, then the a^* count as individuals that are transformationally emergent with respect to D.

Under this analysis, the red and blue particles of Checkers World (section 1.4) are transformationally emergent individuals because they are subject to some laws that are not determined by those that govern the black and white particles. The Checkers World example was constructed so that the laws that are peculiar to the red and blue domains operate in addition to the laws that operate in the white and black domains. There will be other types of systems in which the laws of the new domain override or alter some of the laws of the original domain. The difference between these two types of systems is irrelevant to the emergent status of the individuals in the new domain, but it is central to the discussion when considering downward causation.

A brief clarification of the concept of fundamentality may be helpful here. A standard definition of "fundamentality" within both philosophy and physics is that an entity is fundamental just in case it is not composite. Although not always explicitly stated, it is also often implicitly assumed that the synchronically fundamental entities are immutable. From now on, I shall denote that concept of fundamentality with "fundamental_S" where the subscript_S represents "synchronic"; I will follow a similar convention for "fundamentalism_S"; in that sense, ontological fundamentalism_S is committed to two related aspects of atomism that were discussed in chapter 1: the view that the fundamental_S entities of a domain are indivisible and immutable. Transformational emergence is, therefore, incompatible with any ontology in which all of the fundamental_S entities are atoms, because it violates the immutability condition for atoms. More on this later. Fundamentalism_S need not subscribe to the other aspect of atomism, which is that the atoms are individually distinguishable.

In the case of mobs, D is the domain of individual members of normal society, considered as subject to the laws of individual normal psychology; D^* is the domain of members of a mob, considered as subject to the laws of individual abnormal psychology; and a candidate for $D^{\dagger}$ is the domain of neuroscience, which will subsume both normal and abnormal psychology. Once the members of the mob have undergone a transformation from normal to abnormal individuals, many collective properties—such as the level of violence—will be a function, although probably a nonadditive function, of the levels of violence of the transformed individual members of the mob. That fact means that there is no reason to elevate the collective properties to the status of emergent, rather than aggregative, properties.

Is it defensible to take the abnormal individuals themselves as emergent entities, together with their properties and behavioral laws? Recall our definition of "novelty" from section 1.6: an entity E is novel with respect to a domain D just in case E is not included in the closure of D under closure criteria C that are appropriate for D. The appropriate closure conditions for a domain of concrete entities may vary with the domain, but they will usually include the laws that govern the behavior of the elements of the domain. When the laws change, one is ordinarily in a new domain. The laws governing a domain will not correctly tell us how the transformed individual behaves as a member of a different domain D* unless the laws of D* are a consequence of the laws of D. If D is taken as the domain of individuals whose informal closure conditions exclude individuals with properties that are included in the domain of abnormal psychology, we will have a prima facie reason to consider each abnormal individual in the mob as an essentially different type of individual with at least one novel property that develops from interactions within the original domain. Furthermore, if being an essentially different type of individual entails obeying novel laws, then we have a further prima facie reason for considering the domain of abnormal individuals as emergent. There is no holism involved.

The possibility of including as cases of emergence situations that result from changes in the properties of individuals may seem too generous, too close to cases of ordinary causation to warrant inclusion in the category of emergent properties. Yet most philosophical accounts of causation, dealing as they do with an event ontology, are unconcerned with the kinds of transformations that are the focus here. That is as it should be, since causation is almost always treated as a domain-independent concept for which domain-to-domain transitions are of no concern.[5] Diachronic causal relations can lead to transformation,

5. Specifically, mental causation and physicalist conserved quantities approaches to causation are exceptions to this generalization.

to rearrangements of atoms, to changes in relational properties, and to many other things. Transformational emergence can also be produced by non-causal interactions, of which quantum entanglements are a canonical example.

In these discussions, it is more than usually important to maintain a sharp distinction between theoretical representations of a domain and the elements of the domains themselves. In computer simulations and in other types of models, there are no finer grained features to the primitive elements of the model than have been built into the model, even when what they represent is known to have further structure. We can be certain of this because we constructed the model. These primitive representational elements are subject to no further analysis from the perspective of the model used to describe the domain. For example, in many theoretical treatments, neurons are taken as having no internal structure, giraffes are represented as members of a population and the physiology of individual giraffes is of no concern, economic agents have economic properties but no biological properties, and so on. In reality, the elements of the domain themselves are usually, although not universally, amenable to a finer grained treatment using entities belonging to some other domain, the elements of which may temporally precede the entity that we are investigating. This is one reason claims about emergence in computational models must be clearly separated from claims about emergence in the domain those models are intended to represent.

Transformations mark another important difference between emergence occurring in a computational model and emergence occurring in a natural system. If the code for a computational model results in a reference to an entity that is not included in the syntax for the model, the model will fail to execute properly, so that on a properly debugged computational model, novelty will at best occur within a conceptually enriched enhanced language. This is a further reason computational models are not the right place to look for examples of ontological emergence. There is nothing similar to prevent transformations occurring in natural systems, which is why we should look to those systems for examples of this type of ontological emergence, rather than to surrogates in the modeling world. Representational systems include only what is built into them qua representations.[6] This difference is why ontological emergence must always, at some point, be approached empirically—and with due concern for ontological considerations—rather than purely theoretically. The inferential emergentist position that once the predictive resources of a theory had been exhausted, resort to further observations was necessary in order to establish which laws apply to emergent phenomena reflects this realization, even if the inferentialist approach is too tied to a theoretical approach.

It should be apparent that it is only if the individuals are taken to be basic elements of the domains D and D^* of normal and abnormal human psychology, respectively, and are treated as having no internal structure, as is often done within theoretical models, that the abnormal individuals can be seen as similar to the blue particles of Checkers World. In the case of humans, we have obvious reasons not to do so. In Checkers World, all the particles are synchronically fundamental—they have no further structure and their behavior cannot be further explained beyond appeal to the rules of D and D^*. But, whatever the degree to which behavioral psychology might treat normal and abnormal humans as fundamentally different, in reality they have rich and informative neurological and biochemical structures that offer genuine promise of providing a unifying account for both.

Although not determinative, there is a widespread belief that not only the transformations from normal to abnormal psychology but also the entire domain of abnormal psychology can be explained by moving to the domain, call it $D^{\dagger\dagger}$, of neuroscience and biochemistry. If, within $D^{\dagger\dagger}$, there are unifying laws of normal and abnormal psychology, then it is incorrect to claim that the properties of individual mob members emerge from the properties of normal humans, because the transition from the latter to the former is a consequence of the laws of $D^{\dagger\dagger}$, and hence the properties of abnormal individuals are not novel with respect to $D^{\dagger\dagger}$, which is closed under the production of both normal and abnormal individuals. We can now see why the often-cited case of mobs has a

6. Those who believe there is no representation independent access to the world can reconstrue this point in terms of the need to re-represent systems under pressure from additional empirical facts originating in those systems.

natural appeal as an example of emergence, but it is unlikely to survive as a genuine case of ontological emergence. Viewed from the perspective of everyday classifications, and operating under the default assumption that individual humans are robust units of analysis, apparently the only option for explaining the behavior of mobs is to attribute an emergent holistic property to them. Once the mutability of individuals is recognized, and a moderately scientific account of their behavior is available, we can see that we can recover the generative atomism approach, making the emergent features merely a consequence of our representational schemes. We can also see from this analysis that the anti-emergentist position of methodological individualism rests on the fact that the individuals concerned are not fundamental_S, rather than from individualism itself, a feature that may not be welcomed by many traditional adherents of individualism in the social sciences.[7]

There are social theories that address the phenomenon of smart-crowd formation in a different way from the one that I have just described for mobs. What is known as social identity de-individuation theory asserts that in text-based environments such as communication via text messages, people are seen less as individuals than they are in more traditional visual contexts. They can then form a social identity that trumps their individual identity. What de-individuation emphasizes is a specific kind of loss of identity: the conversion from an identity as individual to an identity as part of a group. This is a different kind of change in the individuals, and it allows the autonomy of population-level laws without claiming that there is downwards causation. There is, perhaps, less reason than in the case of mobs to think that an ontological change in the type of individual has occurred, rather than a shift in the social or psychological classification of that individual, but that is an arguable point. We thus have no unassailable reasons that suggest ontological emergence exists either at the individual or at the collective level.

Despite this, there do exist plausible candidates for transformational emergence.

7. In addition to a dislike of group-level laws, some individualists welcome the idea that humans form a special and irreducible type, sometimes because of emergent properties.

2.2.1 A Possible Example of Transformational Emergence

Because of the rapid pace of change in fundamental physics and the sharp divisions of opinions about what counts as an appropriate fundamental ontology, suggesting candidates for transformational emergence must inevitably be tentative. So, the following example is put forward in the spirit of exploration. If it fails to hold up, we shall gain some understanding either of how the structure and dynamics of the world undermine the prima facie candidacy of this example or of how the process of deciding what is fundamental in physics is underdetermined by the data. An objection to this example might be that the particles terminology is merely shorthand for a field theoretic description. For ease of exposition, I have adopted the particles approach here, although I do not believe that the basic concepts of transformational emergence are restricted to an ontology of individuals.

The Standard Model of physics currently includes different types of particles as fundamental_S. These include three types of quarks and their three corresponding antiparticles, six types of leptons (electrons, muons, taus, and three corresponding types of neutrinos), and force carrier particles (including photons, gluons, positive and negative W bosons, and Z bosons).[8] The Standard Model has an essentially compositional component with respect to nonfundamental particles. Baryons (protons, neutrons, and others) are composed of quarks of various types and mesons are composed of a quark and an antiquark. What is important for us is that, although all six types of leptons are taken to be fundamental in that they are considered to have no components and no internal structure, electrons and neutrinos are stable—in contrast to baryons, they are not composed of quarks—whereas muons and taus are not stable.

We can now formulate a definition of diachronic fundamentality:

> *Definition* A is fundamental_d with respect to a system S just in case A existed at the temporal origin of S. Here, *d* denotes "diachronic."

8. I have not included the Higgs boson in this taxonomy.

Noting that leptons did not exist until after the Big Bang, if we take S as the universe, leptons do not count as fundamental$_d$, even though they are fundamental$_S$. The properties of fundamental$_d$ and fundamental$_S$ are logically independent.

One can then argue that the appearance of a new type of fundamental$_S$ particle counts as an example of emergence under a different, diachronic conception of emergence—that of transformational emergence. One aspect of the Standard Model that is of interest to us is such processes as the transformation of a muon. Muons have various decay modes, the most common of which is the transformation of a muon into an electron, an electron neutrino, and a muon neutrino under the influence of the weak interaction. An important feature for us is that because muons have no constituents, the transformation cannot be represented as the decomposition of the muon. The usual terminology, which I just employed, is one of decay; but that has connotations of radioactive decay in which compositionality plays a role in some cases, whereas in the cosmological case, the process consists in the transformation of one or more types of fundamental$_S$ particle into other types. For example, unstable leptons transform into other (lighter) leptons, neutrinos, quarks, and antiquarks. Quarks can also undergo transformations, such as a charm quark transforming into a strange quark plus a W boson, with the latter then transforming into up and down quarks. These transformations are constrained by conservation laws on electrons, muon, and tau numbers (the lepton number in a given family must remain constant), but the process consists in a transformation and not in the decomposition of a lepton into its constituents, of which it has none. The transformations, because they involve fundamental$_S$ particles that are not fundamental$_d$, do not involve a change in levels, only a change in domain.

These types of processes count as a type of emergence because they have three of the four characteristic features of emergence. As a reminder, these four features are that emergent entities develop from other types of entities, that the emergent entities are autonomous, that the emergent entities are novel, and that some element of holism is involved. Two of the characteristics are straightforward. First, the transformed particles develop from some prior entity or entities. Second, the result of the transformation is an autonomous entity or entities. Once they have appeared, the products of the transformation do not depend on the original particle or particles for their existence. What of novelty? One might argue that the products of a lepton transformation are not novel at the type level because instances of them first appeared during the early period of cosmological development. But as I argued in section 1.6.2, to restrict tokens of emergence to the first ever instance of a type is an overly strict criterion that would rule out almost every case of emergence. So, if we allow subsequent instances of an emergent type, the token entities that occur as a result of lepton transformations are of a novel type. Finally, what the example lacks is the property of holism, but since none of the four characteristics is necessary, that is not a disabling feature of the example. Holism is an artifact of the synchronic approach to emergence.

Still, more needs to be said. The open nature of scientific inquiry requires flagging these examples as tentative. In particular, transformational emergence allows that the laws operating at the initial state of the system did not fix the laws at all times—a position that could be undermined if unification projects show that a single set of underlying laws can account for each of the transformations just noted. The Standard Model could be replaced, in which case we should have a candidate for the domain $D^{\dagger}$ that will explain the appearance of leptons. Indeed, it has been proposed that the period immediately preceding the first appearance of leptons is correctly described by just such a replacement Grand Unified Theory. The view is that at very early cosmological epochs, the strong, weak, and electromagnetic forces are thought to have been unified, only later separating into, first, the electroweak force and the strong force, with the former then splitting into the electromagnetic force and the weak force. However, at present there is no concrete evidence for such a theory, and experimental tests would be extremely difficult. These examples raise the question of whether instances of all the current fundamental particles of the Standard Model existed in the very earliest stages of the universe.

If the simplest form of the Standard Model is correct, then there are actual cases of transformational emergence even in the early universe because the domain of leptons is governed by the conservation of lepton number law, which is different from the conservation of baryon number law. One might say that there is just one law operating throughout, which that the numbers of fundamental particles are conserved. But that looks like a metaphysical principle and not a scientific law. Conservation laws are formulated in terms of some specific property, such as energy, momentum, lepton number, and so on. The diachronic nomological closure of the physical domain is therefore a different commitment from the usual causal closure claim because the latter assumes the laws of fundamental physics are fixed.

Finally, regarding quarks, it is worth noting that quarks are not, in the current cosmological epoch, found in isolation, whereas leptons are. This entails that one criterion for emergence discussed in chapter 1:—that the properties of R(A, B, C) cannot be predicted from the properties of A, B, and C "taken in isolation" could be applied only if we counterfactually, and perhaps counterlegally, assume that the properties of quarks would remain the same were they to be found in isolation.

In contrast to the mob example, the leptons example has a parallel with the Checkers World example of section 1.4, in the sense that within a given system S we have fundamental_S, noncomposite individuals that transform into different noncomposite individuals that are also considered fundamental_S. The new fundamental_S entities do not synchronically depend on anything else, whether nomologically or logically. A case can thus be made that the transformation of one fundamental_S entity into another is sufficiently at odds with the core ideas of generative atomism that it should count as an example of emergence.

I offer a last example that has affinities with the case of mobs, but is different in an important respect. Within the debates about adaptation in biology, one school holds that species are individuals, despite their being composed of tokens of organisms.[9] Within that school, species are considered to be fundamental units of selection, and they change over time, sometimes becoming extinct and replaced by a different species. If natural selection does operate on species rather than on tokens of organisms, perhaps we have a case where the kind of reductionist strategy described for mobs is inappropriate, and one can justifiably claim that one species transformationally emerges from a prior, different species.

9. The position is derivative from Ghiselin (1974) and Hull (1976).

2.3 FUSION EMERGENCE

I have argued that transformational emergence can be produced when at least one essential property of a fundamental_S entity is changed, with an accompanying change in domains. As such, transformational emergence is a generalization of fusion emergence, described in Humphreys (1997a).

The original motivation for drawing attention to the existence of fusion emergence was to address the problem of downward causation.[10] This is a problem that involves finding causal work for emergent properties once one has made a commitment to physicalism and to the causal closure of the physical domain. Consider fundamental_S physicalism. This position is committed to the existence of the entities and properties of fundamental_S physics, and everything else that exists is determined by those entities and properties in virtue of some determination relation D. Suppose now that the realm of fundamental_S physics is causally closed in the sense of section 1.6, and take some event E that occurs after the origin of the universe. If E belongs to the fundamental_S domain, then all the events that causally determine it occur in the domain of fundamental_S physics. So any events outside that domain will be causally redundant with respect to the occurrence of E. If E lies outside the fundamental_S domain, it is, according to fundamental_S physicalism, determined by entities and properties within the fundamental_S domain, perhaps in virtue of some non-causal determination relation such as supervenience. Because E's occurrence is determined by fundamental_S entities, any events outside the fundamental_S domain will again be redundant with respect to E's occurrence. These two cases exhaust the possibilities, and so all events outside the fundamental_S domain are redundant with respect to bringing about E. In particular, it is claimed, emergent properties, which by their nature are not supposed to be fundamental_S, are redundant, and hence are dispensable.

10. Downward causation will be treated in detail in section 3.4.

Somewhat more explicitly for the emergence case, we have the following:

(1) Every physical event E that is caused has a prior sufficient fundamental$_S$ physical cause C.[11]
(2) If an event E has a prior sufficient cause C, then no event C* distinct from C that is not part of the causal chain or process from C to E is causally relevant to E.
(3) The realm of the fundamentally$_S$ physical is causally closed, and so all events in the causal chain or process from C to E are fundamental$_S$ physical events.
(4) No emergent event C* is identical with any fundamental$_S$ physical event.

Therefore,

(5) No emergent event C* is causally relevant to E.

Therefore,

(6) Emergent events are excluded from causally affecting fundamental$_S$ physical events and hence are causally dispensable.

The exclusion argument need not be restricted in its application to fundamental$_S$ physicalism and can be generalized simply by striking the word "fundamental$_S$" in (1), (3), (4) and the conclusion (6) of the argument. It has a much broader scope than simply casting doubt on the causal efficacy of mental properties. If the argument is valid, anything outside the domain of the physical, however one might construe that domain, is causally impotent. In particular, whether one adopts fundamental$_S$ physicalism or a view that applies to the subject matter of physics more broadly, properties that are peculiar to the

11. The argument can be generalized to include probabilistic causes, but I shall not do that here. An early example of the argument can be found in Kim (1993b), in which the target was nonreductive physicalism and the determination relation was one of supervenience.

chemical and biological domains become causally redundant, and hence dispensable.[12]

Although the original role for fusion emergence was to show how supervenience relations fail to apply when fusion occurs, that role is generalizable to any similar dependency relations. Such dependency relations require the continued existence of some basal property instances that can fix the emergent property instance. An essential feature of fusion is the loss of such basal objects or properties when they fuse to produce a unified whole. Having undergone fusion, the original entities no longer exist to provide a base for the dependent properties, and causal pathways are routed through the fused emergent property, which in the cases considered will either be a fundamental$_S$ entity of the original domain or a member of some other domain.[13] In the latter case, the assumption that the fundamental$_S$ domain is causally closed is incorrect, and as a result, premise (3) of the exclusion argument is false. In the former case, there is no downward causation involved and no conflict with the causal closure of the fundamental$_S$ physical domain. In that case, premise (4) is false.

2.3.1 The Money Example: Fusion Without Emergence

In order to ease into the topic, I shall present two examples which, although superficially analogous to fusion emergence, differ in essential ways from it. They involve everyday phenomena that are easy to evaluate. The first example is due to Paul Teller (1992). Consider the exchange of two separate $1 bills for a single $2 bill. Once this exchange has taken place, there is no sense in asking which part of the $2 in your hand is made up of the second bill you handed over. There is simply a fused quantity of two dollars, represented by the single note. If you use the note to pay a bill of $2, it is not the two individual dollars that result in the bill's being paid, but the totality of two dollars. This example does

12. For a detailed examination of, and solution to, the exclusion problem, see Humphreys (1997a).

13. If it is a fundamental$_S$ entity of the original domain, it need not be a new kind of fundamental$_S$ entity. The fusion of existing entities could result in the same type of fundamental$_S$ entity, in which case we should have a case of fusion without emergence.

not involve any emergent properties because no novelty arises as a result of the fusion, but the basic idea about the loss of identity of constituents should be clear. The example also involves a functional property—a dollar can be realized by any number of concrete entities, such as a single banknote, a collection of coins, a specially coded section of computer memory, and so on—and so is not the kind of property usually involved in ontological emergence.

2.3.2 The Probability Example: Failure of Supervenience Without Fusion

The second example is a little more scientific and rests on the fact that the joint probability distribution for two random variables is not fixed by the distributions for the variables taken separately, whereas the joint distribution does determine the marginal distributions. This is easily seen by considering two systems. The first system consists of two fair coins that are probabilistically independent. Then $P(X_1 = H \& X_2 = H) = P(X_1 = H).P(X_2 = H) = 1/4$. The second system consists of two fair coins that are welded together at a point on their circumferences so that they always come up the same side. Then as before $P(X_1 = H) = P(X_2 = H) = ½$, but now $P(X_1 = H \& X_2 = H) = ½ = P(X_1 = T \& X_2 = T)$ and $P(X_1 = H \& X_2 = T) = 0 = P(X_1 = T \& X_2 = H)$. Taking the two examples together, this shows that the probabilities of the individual events alone do not determine the probabilities of the joint events. As a result, the joint distribution does not supervene on the marginal distributions alone. However, if we can specify the dependencies between the marginals, then it is possible to recover the joint distribution. Sklar's theorem is central here. It states that given a joint distribution function on N random variables, having marginal distribution functions $F_1, \ldots, F_N$, there exists a copula function (i.e., a multivariate distribution on the unit hypercube $[0,1]^N$ such that every marginal distribution function is uniform on the interval [0,1]) such that the copula function takes the marginals and returns the joint distribution.[14] As mentioned, the reverse determination does hold—from the joint

14. For details, see Nelsen (1999).

probability distribution, the marginal probability distributions can be determined.

If we consider the probabilities involved to be physical probabilities, then we can correctly say that a determinate probability for the joint distribution does not exist without a specification of the interactions between the systems that generate the individual probabilities. It is also arguable that the marginal probabilities continue to exist within the joint system, and in that way it differs from fusion. This is in contrast to the money example, in which there are functional features of $1 bills that are lost when we have a $2 note, such as when a bus fare of $1 requires the exact amount. The coins example does suggest that the intrinsic properties of the components (here, the marginal probability distributions) are insufficient to determine all properties of the whole system (here, the joint probability distribution) and that the interaction between the components is crucial. In that sense, the joint probability is an example of what I shall call an L-external relation (see section 3.2), since it does not supervene on the intrinsic properties of the relata, but it does supervene on the properties of the composite system.

2.3.3 The Representation of Fusion

We can now use the ideas behind transformational emergence to understand fusion emergence. The basic idea behind fusion, as we have seen, is that two property instances that belong to a domain D interact, and in so doing, the instances are transformed in such a way as to produce a new property instance, the key feature of which is that it does not have the original property instances as components. When the new instance belongs to a different domain D′ and the exclusion problem has to be addressed, the problem disappears because there is no overdetermination between the causal processes in the original domain and those in the new domain. Now we can see why fusion emergence is a special case of transformational emergence. Interactions between entities A and B result in both A and B being transformed in a radical manner, losing the essential property of existence, with the resulting entity being the fusion of the two. The fusion is never a combinatorial rearrangement of existing entities or some other operation that is amenable to GAP. Although I have discussed fusion in terms of examples that involve entities that are fundamental$_S$ to the original domain, this is not necessary. For instance, A and B could be composite and the fusion of the two would then involve the fusion of all or some of their components. In what follows I shall consider only the fundamental$_S$ case.

Here is a formal representation of the situation in which events consist in the instantiation of a property at a spatiotemporal location. Initially, suppose that only one property is instantiated in each event, represented by the notation $P^i_m(x_t)$, which denotes that the property P^i_m is instantiated at the location x at time t. Here, the superscript denotes the domain to which the property belongs. The subscript indexes the properties within that domain, so that P^i_m denotes the mth property in domain i. Again, for simplicity I shall assume that x_t represents a spatiotemporal location, but more general spaces can be used. It might seem that using a spatiotemporal location is a problem in that psychological and other properties in the social sciences are often thought of as lacking spatial locations. But propositional attitudes and other cognitive states are associated with specific individuals who have locations, and economies and cultures are associated with geographic regions even when global, so this objection can be accommodated.

I shall consider the simplest case where the location of the two property instances is the same, denoted by x, and the situation is one in which a change of domain is produced by fusion. If $P_m{}^i(x_t)$, $P^i_n(x_t)$ are i-domain property instances, then the fusion of these two instances, $[P_m{}^i(x_t) * P_n{}^i(x_t)]$ produces a new property instance which is denoted by $[P_m{}^i * P_n{}^i](x_{t'})$ or in an abbreviated form by $P_s{}^j(x_{t'})$. The cases that are most relevant for emergence occur when $j \neq i$ and the fused property belongs to a new domain j. In Humphreys (1997a), I assumed that all fusion resulted in a change in levels or domains. That no longer seems to me to be correct—fusion can lead to a change in domains but need not. I also note that some properties can belong to more than one domain. As an example, chemical properties such as acidity can occur in biological domains, one example being that formic acid occurs in certain kinds of ants. This does not entail that a property can occur in all domains. Although Broad discussed the existence of what he called "ordinally neutral properties," instances of which occur in all domains, this is misleading. A physical property such as mass may be coinstantiated with an economic property such as money—for example, in the case of coins—but that does not put the physical property itself into the economic domain, if only because the mass has no economic function. Money could be realized by massless particles such as polarized photons.

The double use of the asterisk (*) notation should not produce any confusion; the primary use is to represent interactions between property instances and what is fused are properties. Fusion interactions are real features of the world and [.*.] is not a logical or mathematical operation on predicates. The use on predicates is simply a notational device—it does not represent that the properties P_m^i and P_n^i are constituents of the property $[P_m^i * P_n^i]$.

We can then represent the action of fusion, which is usually a dynamic process, by $[P_m^i(x_t) * P_n^i(x_t)] \rightrightarrows P_s^j(x_{t'})$. The nature of the interaction between the original property instances that leads to the fusion will depend on the properties involved. In some cases it might be causal, in others not. Aside from the metaphysical remarks made below, there is no general theory of fusion, and the properties of particular instances of fusion will be determined by the features of particular phenomena. Some fusion interactions will be symmetrical, others not; some will be associative, others not; and so on. The * may exhibit threshold effects, in that it does not operate until F or G reaches a certain value; such a feature requires the use of quantitative properties rather than qualitative properties. In the more general case in which the original properties are instantiated at different locations, the fused instance can be represented by $[P_m^i * P_n^i](z_{t'})$, where $z_{t'}$ is the spatial region occupied at t' by the fused properties. The key feature of $[P_m^i * P_n^i](z_{t'})$ is that it is a unified whole in the sense that, within the fusion $[P_m^i * P_n^i](z_{t'})$, the original property instances no longer exist as distinct entities. In particular, they cannot provide a base for supervening properties.

The result of fusion is not, logically, a relational entity—the fusion operation does not produce a relation between two property instances because there are no longer such instances to relate. Because of this, fusion differs from relational holism as described by Teller (1986): "By *relational holism* I will mean the claim that objects which in at least some circumstances we can identify as separate individuals have *inherent relations*, that is, relations which do not supervene on the non-relational properties of the two individuals" (73). However, it is not relations but dynamic interactions and their outcomes that matter.

A fortiori, fusion is not an instance of an internal relation (see section 3.2). Instead, it is a transformation of two separately existing entities into a single, unified whole. Nor is it a special case of contextualism. Contextualism, despite requiring that the entities are altered in virtue of occurring within a given context, considers the altered entity to maintain a separate existence within the whole, and this is not true of the results of a fusion operation. Fusion is, however, an example of the general feature that in emergent systems the properties of the whole are not determined by the properties of the fundamental entities taken in isolation, a feature that as we have seen (section 1.1.1) was emphasized by Broad.

What of our criteria for emergence? The fused property satisfies the first criterion because it results from an interaction between the original properties. Regarding novelty, the novelty or otherwise of fused entities is not something that fusion itself determines. Although fusion entails the production of a different property from those that enter into the interaction, this does not entail that the property is novel in the sense required for emergence; the fused property could occur within the original domain or in some other domain. So, novelty must be imposed as an additional requirement to have an example of fusion emergence. A holistic element is present in the fact that the fused properties form a whole without identifiable components. Finally, the disappearance of the original property instances renders the fused property autonomous from those original properties.

Some of the principal consequences of the fusion approach are, then, as follows:

(1) Fusion shows that supervenience and other dependency relations are incapable of representing this kind of ontological emergence, because the relevant features no longer exist for the fused feature to depend upon.
(2) There is a dynamic or quasi-dynamic aspect to the fusion process in the sense that fusion results from a process of interaction and it is rarely instantaneous.

(3) The result of fusion is an irreducible, unified, holistic entity.
(4) Emergence can take place within the domain of the physical. This means that although the most fundamental domain is not causally closed, there need be no violations of physical law in virtue of that. In the quantum mechanical examples of fusion emergence discussed below, the same laws of quantum mechanics govern the entangled states as govern the original states.
(5) Fusion itself is insufficient for emergence because the resulting property may lack the required kind of novelty, but when novelty is also present, so is emergence because the other three conditions are always satisfied.

As we saw in section 1.6, many definitions of the causal closure of the physical assert that the physical realm is causally closed just in case every physical event is fixed by some prior physical events and laws—something that I have called "causal completeness." Causal completeness allows that there can be deterministic chains that leave and re-enter the physical domain, and it is compatible with the existence of fusion and defusion. If we replace the definition of causal completeness by the definition of causal closure given in section 1.6, then the burden of proof is on defenders of physicalism to show that the realm they consider to constitute the physical is in fact causally closed in this more precise sense. With a broad construal of the physical that has the upper boundary of the physical located where the realm of the mental begins, the causal closure claim has some plausibility, but it allows for what is usually described as downward causation to occur between various subdomains of the physical. If physicalism is taken to be fundamental physicalism, the causal closure of the realm of the physically fundamental is an empirical claim, not one that is necessarily true, and we should be prepared to find that it is false.

2.3.4 Defusion

We can call the inverse of the fusion operation "defusion." As with fusion, not much that is general can be said about defusion. What results from a defusion process depends upon the method used to defuse the emergent entity. Sometimes the original components will be recovered in defusion, sometimes other entities will result. This is in contrast to, but not incompatible with, an important tradition in atomism, which requires that the analysis of a compound entity always lead to the same unique set of atoms. The analytic process may have different intermediate steps, as when one breaks up a jigsaw puzzle into smaller and smaller elements, but it always arrives at the same set of atomic pieces; and for any case of generative atomism, the end stage of the process should always be the same—a collection of atoms.[15] With a properly constructed formal language, for example, the decomposition process on sentences should be unique, paralleling the generation process. This is the reason we say that the atoms are *the* constituents of the object. In contrast, this unique ultimate decomposability is not true of objects that are not composites of atoms. In the case where there is only a finite number of division steps of a continuous object, many different collections of objects can result from decomposition. Thus, consider a piece of paper ripped by hand into approximate halves, quarters, and so on. No two repetitions of the ripping process will result in exactly the same collection of smallest rippable pieces.

One of the salient features of the defusion operation is that it allows a return to the original domain without involving downward causation. Operations on the fused property that occur entirely within a new domain can result, via defusion, in the appearance of properties in the original domain. This is not downward causation, because the defusion process, at least before it is completed, does not involve causal effects on entities in the original domain; it could well be called "transdomain production." A candidate example is decoherence in quantum mechanics. Interactions between an entangled state and its environment result in the separation of the state into two distinct states in a process that counts as defusion. (For details, see Zurek 2002.) Indeed, it is best to abandon the hierarchical perspective, which originates with synchronic accounts of emergence, and to consider these issues from the perspective of domains.

15. In the case of the jigsaw puzzle, the atoms can all be different. The rule of composition is simple: there must be an exact fit, both spatial and graphical, between two pieces to form a compound piece. In advanced jigsaws, the graphic aspect can be absent.

One final thing. Emergence has always been considered to occur in an "upward" direction, but this is not inevitable. As one possible situation, an infinitely descending series of domains could satisfy the first, second, and third criteria for emergence. Holism would not apply in any direct sense, a fact that reveals how dependent the holism criterion is on a compositional ontology.

An objection to defusion has been raised by Wong (2006). The objection rests on the claim that (a) when fused properties are defused, the result is the original properties, (b) the only reasonable explanation for this is that the original properties were contained in the fusion all along as constituents, and hence (c) this is fatal to a basic feature of fusion, which is that the original properties no longer exist in the fusion. This is an interesting and potentially powerful argument. My original exposition of fusion was misleading on this point, because a schematic diagram used in that paper (Humphreys 1997a) represented the results of a defusion operation as being the same as the property instances that originally fused. As noted earlier, there is no need for this to be true in general, and it will be a contingent fact what the results of a defusion interaction will be.

To see this, consider the money example discussed in section 2.4.1. Take two 10 euro notes and exchange them for a 20 euro note. The two separate property instances of 10 euros have been fused into a single property instance of 20 euros. Now, "defuse" this 20 euro quantity. What will you get? Perhaps two 10 euro notes—but not necessarily. You could well get four 5 euro notes, twenty 1 euro coins, or many other "defusions." So, it is not true that fused properties "naturally" decompose into their original "constituent" properties. Does the 20 euro note contain within it two 10 euro notes? Yes, but only potentially, not actually, and the different potentialities cannot be jointly realized. For if it actually contained the two 10 euro notes, then it also contains four 5 euro notes or twenty 1 euro coins, and so on. If these were actual, I would have invented a method for generating very large sums of money, indeed. So, one might well need to take potentialities seriously as being contained within fusions, a metaphysically important feature brought out by this argument, but this does not mean that the these potentialities are actualized.

Now, consider a different case: casino chips that come in units of \$5. I trade in two \$5 chips for a single \$10 chip. When I trade back in the same casino, I always get two \$5 chips, despite this loss of identity. The explanation for this is in terms of constraints placed by the casino system on transformations. This is why the claim that "fused properties don't lose their identity because they retain the power to decompose into the original constituent properties of the fusion" is incorrect. In a different casino with different rules, I might trade in a \$10 chip for ten \$1 chips, which are of a different type from the original \$5 chips. Yet, the objection is important because it highlights something about quantum states that is not often mentioned. Why do entangled states return measurement values of the same kind that are represented in the basis states? The kind of explanation given for the casino situation does not obviously seem to apply here. So, although we have a response to this objection, it is well worth raising as a general puzzle for fused states.

Finally, consider grains of sand (SiO_2) heated so that they fuse into the pure form of glass. Glass is the amorphous form of SiO_2 and sand is the polycrystalline form of SiO_2. (Here, we have another example where claims of the form that "glass is nothing but SiO_2" are wrong.) Before the fusion, the sand always decomposes into the individual grains of sand, but after the fusion, you cannot predict a priori how it will defuse. It depends, for example, on whether the glass is shattered or melted. Moreover, if the glass is shattered, it can "defuse" in many different ways. And as with the money example, one cannot say, except in a dispositional way, that the glass already contains the products of the shattering.[16] (Note here that the shattering "defusion" can leave us within the same domain as the fused object, whereas with reheating, we can move to a different domain.)

2.3.5 Examples

Two scientific examples can now be used to illustrate fusion emergence. First, and most basic, is the case of quantum entangled states. This

16. If the glass contains weaknesses, it might shatter along those lines, but that does not affect the main point here.

exemplifies the extreme case of failures of immutability and of distinguishability. The case for fusion applying to entangled states has been set out in detail in Kronz and Tiehen (2002).[17] Although fusion emergence was initially described in terms of properties, I kept open the possibility that it applied to other kinds of emergent entities, and one of the virtues of the Kronz and Tiehen article is that it shows that care must be taken to separate properties, states, and objects as candidates for emergence. As a number of authors have correctly noted, entanglement applies to states, and states are where we should look first for fusion emergent features of systems. Kronz and Tiehen also suggest that we distinguish between what they call "radical emergence," of which fusion is an example, and "dynamic emergence." Dynamic emergence holds that "Emergent wholes have contemporaneous parts, but these parts cannot be characterized independently from their respective wholes. Emergent wholes are produced by an essential, ongoing, interaction between [their] parts" (345). For reasons I have given above, I believe that this reference to parts is inadvisable.

A second example, in which the situation is more open to discussion, is covalent bonding in chemistry. The motivation for the idea that "molecules are nothing but collections of atoms" is a compositionality view—molecules are simply structured arrays of atoms. At the token level, this view involves a two-way determination: given the atoms in the appropriate arrangement, one necessarily has the molecule, and given the molecule, one must have the atoms and the arrangement. This compositional view is false for chemical molecules.

Perhaps the simplest example of this involves the hydrogen molecule. When two hydrogen atoms are widely separated, they can each be considered to be composed of a single proton and an associated electron. All four entities in the system of separated atoms are distinguishable from one another—the electrons from the protons in terms of charge value and mass, and the individual electrons and individual protons from one another in virtue of their wide spatial separation. In contrast,

17. Kronz and Tiehen argue that a more general form of emergence is common in quantum mechanics and that a more squarely dynamical form of emergence is preferable to fusion. See also Hutteman (2005) for a response to Kronz and Tiehen.

when the atoms are spatially proximate to one another, because of the indistinguishability of electrons, the electron wave functions overlap to such an extent that it is impossible to individuate them. Although the standard treatments usually talk of two indistinguishable electrons, what one really has is a joint probability distribution within which there is no sense to be made of separate, coexisting particles.[18] This means that if the original, spatially separate hydrogen atom 1 is identified with proton 1 and electron 1, and the original, spatially separate hydrogen atom 2 is identified with proton 2 and electron 2, within the bonded molecule it is no longer possible to say that hydrogen atom 1 (or 2) exists as an identifiable subunit.

When covalent bonding is present, it is therefore incorrect to think of molecules as being composed of individual atoms bonded together; one must think in terms of a set of nucleons together with a charge-density distribution.[19] It is not that there are two electrons and one cannot say which belongs to which atom. It is that it no longer makes sense, given the overwhelming spatial overlap of the wave functions, to consider the electrons as anything but a fused entity—a single "object" that is shared by the two protons (which retain their separate identities). Because this results in a failure of the distinguishability principle, generative atomism fails even for the simple case of hydrogen atoms. When the outer electrons of two atoms produce a covalent bond, the charge and mass of the nucleons remain unchanged, as does the total charge of the molecule. What does change is a slight lowering of the energy of the combined molecular arrangement compared to the energies of the atoms before the interaction and it is this change in energy that is responsible for the novel properties of the molecule compared with the original atoms.[20]

What I have just described is a standard interpretation of valence bond theory. More generally, the molecular orbital method considers all

18. See, for example, Eisberg and Resnick (1985, 421).

19. This is the standard view in contemporary chemistry. "Organic chemists do not think of molecules as being made up only of atoms, however. We often envision them as collections of nuclei and electrons, and we consider the electrons to be constrained to certain regions of space (orbitals) around the nuclei" (Carroll 1997, 4).

20. The first workable theory of covalent bonding was given by Heitler and London in 1927—something of an *annus mirabilus* in quantum theory (see Masterton and

the electrons to belong to the molecule as a whole, rather than just the pair of outermost electrons, but in that case, too, the distinguishability of the original electrons and atoms has been lost. Quantum mechanical calculations show that having the electron density of the outermost electrons spread out over the entire molecule, rather than being concentrated in the regions around individual atoms, decreases the overall energy and enhances the stability of the molecule. However, these theoretical treatments of chemical bonding are tightly tied to models and to the use of computational approximations, making conclusions about ontological emergence indirect.[21] It also has to be mentioned that the status of various kinds of bonds, and even their existence, is not a settled matter among chemists. There is considerable controversy about how many types of bonds exist, and even whether chemical bonds are a legitimate concept, but we do not need to resolve those disputes here. What is curious about the example of covalent bonding is that it is specifically quantum mechanical features that underpin the claim of emergence, and so the very advance that once led to the rejection of chemistry as a source of emergent phenomena now gives us reason to think that emergence is common at the molecular level. The claim is not that all chemical phenomena are emergent—ionic bonding, for example, is not—but that one of the most important features of chemistry gives rise to phenomena that can be considered emergent.

Compare this case to an everyday example of something that initially looks similar. We take two lumps of putty, knead them together, and we have a lump that has twice the mass of the originals. There is a sense in which we have lost the original lumps of putty qua lumps, but we believe that we could identify the parts of the two original lumps, for example, by coloring the original lumps in distinctive shades and that, despite epistemological problems in tracking the parts, the mixture is in

Slowinski 1969). Gilbert Lewis in 1916 was the first to suggest that electron pairs were shared between atoms. Molecules are the result of covalent bonds, not ionic bonds, which are the result of electrostatic forces between positively and negatively charged ions—that is, atoms that have gained or lost an electron from their usual electrically neutral state. The difference between polar and pure covalent bonds is not relevant here.

21. For a discussion of the effects that different models can have on conclusions about reduction, see Hendry (2006, 181–85).

fact composed of the original two masses. The belief is that at the atomic level, the atoms of each lump are preserved and the original "left-hand lump" is present as a scattered object within the whole. If we allow that scattered objects exist, then the lumps are still there—not qua lumps, but scattered around the jointly occupied volume. Lumps of putty are, of course, not atoms of anything—they are neither immutable nor indivisible. Yet these examples indicate something important: generative atomistic physicalism, or GAP, must frequently appeal to reductionist strategies in order to retain its mereological basis. And that in turn seems to mean that if you want to be a physicalist of the GAP, then you must make your physicalist commitments at a level at or very close to what is currently considered to be the fundamental one. In turn, if generative atomism fails at this level of ontological commitment, as it appears to do, then GAP's strategic line of retreat is cut off.[22]

These two examples of fusion emergence have been investigated for decades, and the interpretations given here are widely accepted by scientists. So, fusion emergence is localized but common, not because of some exotic phenomenon occurring deep in a Swiss accelerator, but in almost every macroscopic object that we encounter.

As a nonphysical example, consider a variant on our earlier example of mobs. Take two mobs that coalesce, where the mobs are similar in the sense that they do not differ in ideology, intent, and type composition.[23] The identities of the original mobs are lost when they merge, and it is not possible to say that original mob 1 is here and original mob 2 is there within the whole. Now, suppose that each original mob had subcritical size—that is, that it was unruly but not violent—but that the fusion of the two produces a larger, violent mob. Then, we have fusion emergence at the mob level owing to the novel property that appears. Of course, given the arguments of section 2.2, if we allow a treatment at the level of individuals, the emergence will likely turn out to be illusory, but if

22. This leaves open the possibility that there is a mereological relation between quarks and these larger objects, perhaps one of mereological supervenience, but that relation would have to skip at least one level, and would fail to provide us with an explanation of such things as the exchange forces offered by contemporary accounts.

23. This condition excludes cases such as two mobs of soccer hooligans supporting rival teams—mobs in which the members wear identifying clothing, and so on.

groups of individuals are the units of analysis, then with respect to those units, we have a case of emergence.

Finally, as a more complex example, one can consider the Coalition Government in the United Kingdom during the war years 1940–45. Formed as a result of the collapse of Neville Chamberlain's leadership, the wartime cabinet contained roughly equal numbers of Conservative and Labour politicians, with the occasional National or Independent member. Although these individuals did not, of course, lose their personal identities in becoming members of that group, nor did they abandon their preexisting political beliefs, those beliefs were, under the exigencies of wartime, largely subjugated in favor of a common goal, with the result that previously separate parties fused into a new group with novel features that were peculiar to that coalition.[24] The extent to which interactions between cabinet members who, in ordinary circumstances would not have been colleagues, produced novel solutions to problems is a task best left to professional historians, but this example is an intriguing candidate for fusion emergence in the political domain. The fusion is between two previously distinct political parties, with the possible added element of politically essential changes to the individuals making up the group. This example rests on treating the individuals as fundamental units of analysis.

2.4 ARGUMENTS AGAINST FUSION EMERGENCE

In his paper "Emergents from Fusion," Hong Yu Wong (2006) argues that the fusion account of emergence is unmotivated as a solution to the exclusion problem and is affected by internal problems. Wong has three principal arguments against fusion. These arguments are valuable because Wong's approach throws into sharp relief the clear division between those who emphasize compositional ontologies and those, such as myself, who have argued that such compositional ontologies are

24. Although, as a result of the beginnings of the Great Depression, there had been a National Government in the United Kingdom from 1931 onward, many members of the mainstream parties did not participate. The 1940 coalition was a far more cooperative arrangement.

inadequate for scientific purposes.[25] Although, as I shall show, Wong's arguments do identify constraints that exist on the types of systems that can exhibit fusion emergence, and thus are a helpful supplement to the fusion account, they do not show that fusion emergence cannot exist or that it is a rare phenomenon.

1. *The first argument: Fusion emergence destroys structural properties.* Wong's concern is that fusion will destroy structural properties that are necessary for a system to survive as a structured entity. He argues in this way:

> Consider a system *s* with emergent property *E*. The basal properties giving rise to *E* also constitute myriad non-emergent, structural properties of *s*. If these lower level properties literally ceased to be in fusing into *E*, then so, it seems, would those structural properties. These structural properties may include those crucial to the proper functioning of the system. (2006, 355)

This is what Wong calls the problem of "collapsing structural properties" and says "the collapse of structural properties threatens any empirical situation where basal properties are involved in multiple functions. The problems arise from a neglect of structural properties in fusion emergentism" (356). The problem is supposed to be extreme, in that "on Humphreys's metaphysics, when we have mental properties we have *no* neurophysiological properties—or at least those neurophysiological properties that we take to be correlates of mentality will *not* be co-present" (356).

By way of a response, I note that most systems possess multiple properties, some of which are essential to carrying out the system's function, whereas others are not. A tabletop cannot completely lose its properties of rigidity and flatness while still serving the function of a table, but it can change its color and size and still be a table. When the outer electrons of two atoms fuse to produce a molecular bond, this does not alter the total charge. In general, when a system's state

25. Wong is not a defender of compositional ontologies in the case of properties.

is given by <P, Q, R, . . . Z> (x) and P(x) fuses with Q(x), the properties R, . . . , Z will remain unchanged. In other situations, part of the quantitative property P and part of the quantitative property Q will fuse, leaving the remainder to maintain the state. This can happen when there are surplus quantities of various properties; many processes require only that a threshold level of some quantity is present, the rest playing no causal role. Wong does discuss the case of multiple properties (356–57, especially n. 357): "The parts of the original complex property that remain can then play a completely different, and purely basal, causal role from the emergent property" and adds "the challenge to Humphreys is to demonstrate that this will *always* be the case."[26] But, it is only necessary that there be some examples of fusion emergence in order to show that our world's ontology is not wholly compositional. If, as a matter of fact, the sustaining properties for a system go out of existence when fusion occurs, then either emergence will not occur because there will be no system for the properties to attach to or the emergent property will be instantiated by a different object. In this sense, Wong's objection forms an important constraint on when fusion emergence can occur.

2. *The Second Argument.* The second argument against fusion is connected with the first: we frequently discover properties at a lower level that are associated with the higher level properties. This point is correct, and the fact that scientists can often identify such associated states is a major driving force behind physicalism. But in order for the point to serve as an argument against fusion emergence, this situation has to hold universally. It is a matter for empirical discovery whether fundamental$_S$ physical states are associated with every physical and special science state. The current evidence suggests that fusion and transformational emergence are not uncommon in the physical and chemical domains, but it is far from being ubiquitous. The examples also suggest that the realm of the mental is currently not the best place to look for examples of emergence. Wong is correct that the

26. See also O'Connor and Wong (2005, 665), where they note that because an emergent state will have to persist over time and change, "we cannot equate it with the total physical state of the system at any given time."

evidence from neuroscience does not suggest the existence of autonomous, fusion emergent mental states and properties, but no part of the fusion emergence account suggests that such things should exist. Ontological emergence is where the world puts it; the fact that most philosophers find brains and beliefs more interesting than barium and bonds is a separate matter.

3. *The Third Argument: Basal Loss is Unmotivated.* The third objection, unlike the first two, is misplaced. Wong claims that: "Standard accounts of supervenience are not committed to over-determination, because the basal and supervenient properties are not distinct, as the base (at least) nomologically necessitates the supervenient and they are even spatio-temporally coincident" (358). In cases where the supervenient feature is a redescription of the basal feature, this is correct; but where the necessitation in supervenience is nomological necessitation, the supervening event will usually be distinct from its base. He does correctly note that in order for overdetermination to occur, the two candidates causing the effect must be capable of producing exactly the same effect, for otherwise we would not be in a situation of the very same event being overdetermined (358).

This is an interesting point, but it has the dialectic of the argument reversed. Fusion emergent properties are not in causal competition with basal properties, except where an alternative causal influence happens, by coincidence, to be present at the basal level. Fusion emergent properties can operate entirely within their own domain, without ever directly affecting basal properties. Moreover, a property can have more than one type of effect. The electrical charge of a moving particle can produce both a Coulomb force (an attractive or repulsive electrical force) and a magnetic force. The claim that "If basal properties don't possess the causal powers of emergents, then they can't cause the same effects; so they can't compete as overdeterminers" must then be clarified in the light of these multiple possible effects. Using a version of the causal criterion for property identity (see Shoemaker 1980), the basal properties cannot possess all and only the causal powers of the emergent property, but if they have some, but not all, such causal powers in common with the emergent property, then they can compete as overdeterminers without thereby being identical with it.

In the course of an interesting historical study, Jonardon Ganeri (2011) has also argued that fusion emergence is flawed. Although his reasons fall along the lines of Wong, augmented by a desire to preserve emergence as a supervenience dependent position, the view that he prefers merits attention because in some ways it is sympathetic to the more general transformational emergence. He argues that the ancient Indian philosopher Bṛhaspati (who lived prior to the 6th century CE, dates unknown) anticipated supervenience accounts of the mental and certain aspects of the British emergentists' positions, and that later Indian philosophers argued in favor of a "transformational" account of emergence.[27] The key element of this transformational view is:

> Let us suppose that the blending or combining of the elements "transforms" them in such a way that in their transformed state their combination, according to homopathic principles, instantiates psychological properties. Then it will be true to say that mental properties are reducible to the properties of the transformed physical base but equally true that they are irreducible to the properties of the untransformed base. (Ganeri 2011, 686)

This move, important as it is in rejecting one of the core tenets of generative atomism, is not yet enough to get us to transformational emergence, because Ganeri goes on to say that:

> the idea seems to be that the elements themselves acquire new causal powers when they are in a certain state, namely the state of jointly composing a body, powers that they did not have beforehand when they were in other combinations with other elements. It is . . . the view that the parts themselves have new powers *conditionally* upon their membership of the whole. (Ganeri 2011, 686)

In contrast, transformational emergence does not require that the transformed entities be members of larger wholes, as we have seen with

27. Thanks to Alexandre Guay and Olivier Sartenaer for bringing Ganeri's paper to my attention.

the examples in section 2.2. Dynamic interactions with other entities will do, as long as the changes in the objects are genuinely novel.

2.5 THE ORIGINS OF THE UNIVERSE ARGUMENT

Universal determinism often puzzles philosophical novices because it is committed to the view that the present state of the universe was fixed millions of years ago.[28] The puzzlement can be made to go away, but there is a related puzzle that is less easy to resolve: How can the relevant information needed to fix all future states of the universe be encoded into a single, instantaneous prior state? This puzzle becomes more acute when we consider the state at the first instant of the universe. How can everything needed to guide the development of the universe for evermore be right there in a single time slice which has no history and no lawlike regularities to direct its subsequent development? This is a serious difficulty for any view which insists that there can be no causal properties without an associated law, and that regularities are a necessary condition of having a law. On that view, there are no nontrivial regularities at the start of the universe, hence no laws and no causes. Fundamental$_d$ physicalism would be restricted to what existed at that first instant, and a complete theory would be a complete theory of the physics of that first instant.

There is an easy enough answer to be found by analogy to algorithms, you might say. Every state of an algorithmic process is automatically generated from a single initial state, and we do not find this at all odd. There is, therefore, no difficulty for systems that can refer to rules. But the analogy is unconvincing. Despite attempts to argue for digital physics—the position that the universe computationally processes information—there is a lack of empirical evidence for this view. Even if we assume that, from the beginning of the universe, algorithmic laws of nature existed separately from the specific initial state—off to the side, as it were—and guiding the development of the universe, we are thereby

28. Parts of this argument were first presented in Humphreys (1995).

committed to nomic generalizations existing as a separate category from concrete particulars, and in some sense constraining and directing the dynamics. This is a tenable position, but it requires the holder to explain how this influence from the laws to the specific states takes place.[29]

There are three principal approaches to addressing this puzzle. The first approach says simply that there is no answer to our initial question, that there are just successive states of the universe, one after another, and parts of the first time slice can be attributed a causal role only if and when regularities appear. A problem with this approach is that many events in the initial state of the universe (and others that occurred before a regular pattern established itself) would have changed their status from not being a cause to being a cause when a regularity was later in place. This is not a fatal objection to the approach, because in indeterministic worlds, to which some counterfactual and some probabilistic theories of causation apply, one has to wait for the effect event to occur before it becomes fixed whether an earlier event was a cause of that effect.

The second approach denies an implicit assumption of the puzzle, which is that at the first instant no later instants are yet in existence, and thereby adopts the block universe model with its assumption that all states are eternally coexistent. This second approach has a defect similar to the one we noted about the first approach. Take a given initial segment of the universe. Standard underdetermination arguments show that there are different regularities compatible with this segment and so different subsequent histories of the universe would give rise to different laws. Because explanations of why later states of the universe have the features that they do depend upon both the initial state and the laws, it is not true that any initial segment of the universe determined

29. A short clarification of one point is needed. I put the argument in terms of causal properties rather than causal relations because, if there are no causal relations between simultaneous events, there can be no causal relations at the first instant of the universe—there is only one event or multiple simultaneous events at that point. In contrast, interactions can occur between entities at a single time point. The point also holds for law-based positions. For those accounts, some event has a causal property only if the event is relevantly similar to another event which appears as part of a suitable regularity, and that regularity involves at least the relation of being succeeded by a suitable effect-type event and all the relevantly similar events in the regularity also possess that property.

(by itself) what was to come, but only the entire subsequent development of the universe. This requires a massive degree of temporal holism for the account to work. Causation is no longer a local business.

This is a high price to pay, and one that can be avoided by a third approach, which is to reject the idea that laws played a role in the development of the universe, and to assert that it was temporally local property instances and their relations that alone gave rise to future developments. What the origins of the universe argument shows is that it is the dispositional properties of the initial time slice—those that intrinsically determined "where the universe goes next"—and the interactions between the initial properties that should be of most direct interest to singular causal theorists.

3

Ancestors and Relatives

3.1 J. S. MILL AND C. D. BROAD ON EMERGENCE

Having given an account of transformational and fusion emergence in the previous chapter, it will help to provide a brief historical survey of some earlier approaches to emergence so that we can place those accounts within a larger tradition. The aim is to show that transformational and fusion emergence are more than just additions to a growing list of candidates for emergence; they capture a principled approach that has been present since the earliest explicit investigations into emergence. This is not intended to be a scholarly survey—the philosophers discussed here had subtle and complicated positions, and there are many aspects of their positions that I shall not discuss. These earlier writers neither had access to the wealth of examples that we have today nor were they able to use modern techniques for dealing with nonlinear systems and for representing limits on predictability. They were trading in the zone of the unknown, never a good place to do philosophy. For example, Samuel Alexander represented reality as consisting in a set of levels that, from the bottom, were Space-Time, primary qualities, secondary qualities, life, mind, and at the very top, Deity.[1] Recalling that in the 1920s the nature of phenomenal states was not well understood and that the era of molecular biology began three decades later, it is not surprising that an impenetrable fog hung over the higher elevations of Alexander's six levels.

One might have expected that as the fog cleared, it would have revealed the realm of the emergent to be empty. In fact, as we have seen, although the discovery of quantum mechanical explanations for

1. Alexander (1920).

chemical bonding and the discoveries in molecular biology that led to insights into the nature of life did initially make appeals to emergence implausible or unnecessary, the situation is now different. We have discussed a number of plausible candidates for ontological emergence, and molecular biology contains examples of inferential emergence, such as a current inability to precisely predict the dynamics of many protein folding processes.[2] It is thus a curious fact that increased scientific knowledge has not expelled emergence from the realm of well-understood science; instead, it has provided solid evidence that emergent phenomena lie within the various natural sciences. This fact suggests that drawing philosophical morals from the apparent failure of the British Emergentist tradition should be done with great care.[3]

A second reason for caution about this period is that although some of the writers, such as Mill and Broad, are admirably clear—Nietzsche memorably condemned Mill for his "offensive clarity"—others, such as Samuel Alexander, are far from transparent and obscurity is an inexhaustible source of academic controversy.[4] Indeed, one aspect of British Emergentism that makes it so unappealing is encapsulated in Samuel Alexander's claim about emergence that "It admits no explanation" (Alexander 1920, 2: 47), a defeatist view that is best avoided. I shall therefore concentrate on certain aspects of Mill's and Broad's positions—philosophy that was done in the light rather than in the shade. In doing so, I shall emphasize a different aspect of their positions than does Brian McLaughlin (1992) in his excellent survey of the British Emergentist tradition.[5] There, configurational forces loom large; here, I shall focus on invariance conditions.

2. Dill et al. (2008).
3. McLaughlin (1992) is very clear, and also correct, that the British Emergentist movement was abandoned because of scientific advances, not because of internal philosophical deficiencies.
4. Nietzsche (1889/1968, 67).
5. For additional historical information, see Timothy O'Connor and Hong Yu Wong, "Emergent Properties," in the *Stanford Encyclopedia of Philosophy*, http://plato.stanford.edu/entries/properties-emergent/. Stephan (1992) discusses some post-1925 aspects of the literature on emergence. My conclusions are restricted to the positions of Mill and Broad, rather than to British Emergentism as a whole. "Movements" such as British Emergentism can be historians' constructions of convenience.

We can begin with C. D. Broad's characterization of Pure Mechanism. Although Broad's own views fall into the inferential approach to emergence, we shall draw some ontological analogies from his work later in this chapter. Broad had this to say about Pure Mechanism:

> On a purely mechanical theory all the apparently different kinds of matter would be made of the same stuff. They would differ only in the number, arrangement and movements of their constituent particles. And their apparently different kinds of behaviour would not be ultimately different. For they would all be deducible by a simple principle of composition from the mutual influences of the particles taken in pairs; and these mutual influences would all obey a single law *which is quite independent of the configuration and surroundings in which the particles happen to find themselves.* The ideal which we have been describing may be called "Pure Mechanism." (Broad 1925, 45–46, emphasis added)

Furthermore, "Each particle of this [material] obeys one elementary law of behaviour, and *continues to do so no matter how complex may be the collection of particles of which it is a constituent*" (76, emphasis added).[6]

We see here three things. The first is that Pure Mechanism is a particularly clear example of a generative atomist position. The second is that the compositional principles involve only pairwise interactions. As Butterfield (2012, 105) has pointed out, configurational forces are those that are not determined by the pairwise interactions between the

6. Broad (1925) continues: "There is one uniform law of composition, connecting the behaviour of groups of these particles as wholes with the behaviour which each would show in isolation and with the structure of the group. All the apparently different kinds of stuff are just differently arranged groups of different numbers of the one kind of elementary particle; and all the apparently peculiar laws of behaviour are simply special cases which could be deduced in theory from the structure of the whole under consideration, the one elementary law of behaviour for isolated particles, and the one universal law of composition. On such a view the external world has the greatest amount of unity which is conceivable. There is really only one science, and the various 'special sciences' are just particular cases of it" (76; also quoted in O'Connor and Wong 2005 and McLaughlin 1997, 31). Note the use of "special sciences," almost fifty years before Fodor's (1974) influential arguments for the irreducibility of some of those sciences.

elements of the system. The third is Broad's stipulation that in these mechanistic systems an invariance condition holds, one that requires the laws governing the constituents of a system to operate in a context-independent manner. It is the failure of this and related invariance conditions that is characteristic of many systems that produce candidates for ontologically emergent features.

In Broad's characterization, what you need for a mechanistic account is, first, the possibility of knowing the behavior of the whole system on the basis of knowledge of the pairwise interactions of its parts; and second, knowledge that those pairwise interactions between the parts remain invariant within other combinations.[7] If these two items of knowledge give you knowledge of the features of the whole system, then it is not an emergent system.

Broad placed a great deal of emphasis on what he called "transordinal laws" which connect levels. The transordinal laws are emergent when they are not deducible (explainable) from the laws of lower orders. As he put it, "The only peculiarity of [a transordinal law] is that we must wait till we meet with an actual instance of an object of the higher order before we can discover such a law; and that we cannot possibly deduce it beforehand from any combination of laws which we have discovered by observing aggregates of a lower order" (1925, 79).[8] So, if we are willing to allow that levels are determined in part by the number of constituents in a system, and that at a certain point, the degree of complexity of

7. According to Huttemann and Terzidis (2000, 271), a property of a compound system is mechanistic if: (a) the components and their arrangement determine the property of the compound system; and (b) it is—at least in principle—possible to deduce or explain the property on the basis of (i) the properties of the isolated components, (ii) general laws of combination, and (iii) general laws of interaction.

8. C. D. Broad's account of emergence was also, without using the term, one that relied on supervenience: "No doubt the properties of silver-chloride are completely *determined* by those of silver and of chlorine; in the sense that whenever you have a whole composed of these two elements in certain proportions and relations you have something with the characteristic properties of silver-chloride" (1925, 64). There is a difference: with supervenience, for which the determination relations between the physical and chemical levels are neither explainable nor explanatory (see Horgan 1993 for the latter point). Broad's transordinal laws are not themselves open to explanation, but once discovered by induction, they can serve as the basis of explanations.

interactions between constituents results in the impossibility of deducing the behavior of systems in the next level from the laws that describe the behavior of lower level systems, then we have a case of emergence in Broad's sense.

Although much attention has been paid to the role of complexity and transordinal laws in Broad's approach, the failure of invariance plays at least as great a role in his account of emergence, as we can see from the claim cited in chapter 1:

> The emergent theory asserts that there are certain wholes, composed (say) of constituents A, B, and C in a relation R to each other, that all wholes composed of constituents of the same kind as A, B, and C in relations of the same kind as R have certain characteristic properties . . . and that the characteristic properties of the whole R(A, B, C) cannot, even in theory, be deduced from the most complete knowledge of the properties of A, B, and C in isolation or in other wholes which are not of the form R(A, B, C). (Broad 1925, 61)

So, in the example considered, the particular structure R(A, B, C) is sufficiently novel that information about different structures containing some of, all of, or more than the components A, B, and C is insufficient to allow us to deduce the properties of R(A, B, C). This conflicts with generative atomism because there is a whole for which there is no rule composing its properties from those of its parts, nor is there an analytic procedure that allows decomposing larger composite entities to arrive at the properties of R(A, B, C). Of course, the example is purely for illustrative purposes and the underivable structure will usually be far more complex than a ternary relation.

One can understand Broad's approach by thinking in terms of a simplistic model of atoms and molecules. Suppose we have a hydrogen atom h in isolation. We then combine it with another hydrogen atom using the relation $R_2(h,h)$ to form a hydrogen molecule. We shall suppose that the properties of hydrogen molecules are either predictable from the components of their constituents or are known from experience. We also have an oxygen atom o in isolation. Now, combine the two hydrogen atoms and the oxygen atom using relation R_3 to obtain $R_3(h,h,o)$. Broad's claim is then that from a complete knowledge of the properties of h and o in isolation, and of h in the relation $R_2(h,h)$, and indeed of h and of o in any relational contexts other than R_3, it is impossible to predict the properties of $R_3(h,h,o)$. It is not necessary that the other contexts be simpler; we do happen to have a theory that tells us why $R_5(h,h,h,h,o)$ is unstable that also covers the $R_3(h,h,o)$ case, but that depends on the nature of the subject matter. It is also incompatible with the spirit of generative atomism if we cannot take a larger unit and by analysis decompose it so as to arrive at the properties of $R_3(h,h,o)$. If there is at least one property of $R_3(h,h,o)$ that it is not possible to deduce given a knowledge of those situations, then that nondeducible property is emergent.

This simplistic model is completely unrepresentative of our contemporary knowledge of how atoms combine to form molecules, but it does provide a starting point from which to see how the failure of generative atomism leads to emergence in Broad's position. For him, what leads to the failure of predictability is the fact that the law statements concerning combinations in the configuration given by R_3 do not follow a priori from the law statements about combination in other configurations, but must be learned from experience. Note that the inability to predict is not restricted to the inability to predict laws involving three or more constituents from those involving only pairwise interactions. If the behavior of an entity interacting with seventeen other entities was unpredictable from its behavior when interacting with sixteen other entities, we should still have a case falling under Broad's characterization.

An ancestor of Broad's position on invariance can be found in John Stuart Mill's account of heteropathic effects, one of the first explicit discussions of emergence, although not under that name (Mill 1843, book III, chap. 6, section 1).[9] For Mill, violations of the principle of

9. G. H. Lewes introduced the term "emergence" in his *Problems of Life and Mind* (1874, vol. II, prob. V, chap. iii, 412). For Lewes, an emergent feature is simply an effect of a heteropathic law. Ganeri (2011) makes the interesting claim that an important influence on Mill's views may have been Henry T. Colebrooke, with whom Mill corresponded, and whose 1837 lecture (Colebrook 1837) contains ideas, based on ancient Indian philosophers, that can be construed as precursors of some themes found in the British Emergentist writers.

the composition of causes result in properties that are subject to emergent or, as Mill called them, "heteropathic laws." The parallelogram of forces is a core example of the composition of causes—the combined effect of two forces acting in the "mechanical mode" results from the vector sum of the two forces acting alone. Although much attention has focused on whether emergent processes violate an "additivity" principle for causes and forces, and Mill does repeatedly refer to the sums of various influences, what is of central importance to Mill's account is that the constituents of compound systems—whether they be laws, causes, forces, or some other entity—remain invariant in their effects whether they occur in isolation or as part of a larger whole. The composition-of-causes condition is, therefore, nothing more than an immutability condition applied to causal components. Because Mill considered the issue of heteropathic laws as a causal issue, one cannot separate his views on this topic from his general theory of causation, and for Mill, a factor is required to act invariantly in order to count as a cause.

Why was this immutability condition imposed on causes? For Mill, it was a requirement for a cause that it be invariant in its effects, under the prevailing assumption that all causes were sufficient for their effects. If a purported causal factor C produced its effect in one context but not in another, there must be at least one factor that needs to be added to C to accurately identify the actual sufficient cause.[10] The indivisibility of atomistic causes is a less straightforward issue because many causal factors are continuous, although Mill himself represents causal factors as discrete units. Such discrete, sufficient, causal factors cannot be split without losing their sufficiency and hence, for Mill, their causal status. Regarding the distinguishability feature of generative atomism, the experimental methods of causal discovery advocated by Mill allow us to analytically distinguish individual causal components, and when the immutability condition is satisfied, we can recombine them using the generative apparatus. This invariance also provides the external validity for causal factors that is so prized in social systems—the ability to extend

10. Although Mill's view was formulated under the assumption of determinism, it can be extended to include indeterministic causation. The arguments for this are given in Humphreys (1989, section 25).

findings from experimental contexts to nonexperimental contexts. In fact, what is actually at issue is not experimental and nonexperimental contexts, but the broader ability to embed a causal factor within a new set of causal factors where the original factor continues to act invariantly. In the case of laws, Mill argues that in order to make the effects of the whole deducible from the components, "it is only necessary that the same law which expresses the effect of each cause acting by itself, shall also correctly express the part due to that cause, of the effect which follows from the two together" (1843, book III, chap. VI, section 1).

When the immutability of laws holds, we have homopathic laws; when it fails, we have heteropathic laws. Characteristic examples of the effects of heteropathic laws, according to both Mill and Broad, could be found in both chemistry and biology. As Mill put it, "Not a trace of the properties of hydrogen or of oxygen is observable in those of their compound, water" and "[In the science of chemistry] most of the uniformities to which the causes conformed when separate cease altogether when they are conjoined; and we are not, at least in the present state of our knowledge, able to foresee what result will follow from any new combination, until we have tried the specific experiment" (chap. VI, section 1). One sees here how much Broad was indebted to Mill, as well as the epistemological hedge regarding "the present state of knowledge," later to become a central appeal in Hempel's views on emergence. Mill's view was not restricted to chemistry. He also argued that the characteristic properties of living organisms could never be arrived at by "summing up" the properties of their physical constituents (chap. VI, section 1).

In generalizing these points, three things need to be available to make a deductive inference from the effects that components have in isolation to the effects that those components have when acting jointly. One is a rule telling us how those components combine to form a joint system; the second is the values the components have within the joint system; and the third is the ability to make the deduction within the constraints imposed by the available theory. If the values of the components remain constant across the presence of other factors, we have the second piece of required information. If not, then further empirical information is needed to ascertain the values that occur within a larger whole. Regarding the first item, there is no need to require that the rule of combination have an additive form. Any computable function of the inputs will suffice for the deductions to be made, at least in principle. The form of the combination rule will have to be learned a posteriori, but that is true even when the form is additive. The "additivity" of factors is thus a red herring if one is interested in the difference between systems that produce non-emergent rather than emergent features.[11]

Regarding the last requirement the fact that transordinal laws cannot be deduced from other laws, and knowledge of actual instances must be discovered, undermines the very idea of a deductively closed set of principles. Laplace's (1902) superintelligent agent, famously able to predict all future states of a classical mechanical universe given the laws and the current state of the universe, would not be able to predict an emergent phenomenon without, *per impossibile,* prior access to transordinal laws. This is one reason the insistence that physics is a causally or deductively closed domain is so important to physicalists. Inferential emergence can result from a violation of the deductive closure of a domain, whereas ontological emergence can result from the violation of causal closure.

The most straightforward reason that failures of prediction occur is that the complexity of interactions between the factors at some point surpasses our theoretical and computational abilities to use the factors for predictions. Extending this epistemic point to the claim that sheer complexity might give rise to novel properties is less plausible. We are all familiar with threshold phenomena for which there exist points beyond which qualitatively different effects are produced. For instance, dams bursting when the water pressure reaches a certain level, capacitors firing when the charge reaches a given level, prisoners cracking when the pain from torture passes beyond all endurance; each of these involves a threshold effect. Whether these examples are in any way analogous to complexity thresholds can only be answered by a detailed examination of specific examples of emergence.

11. McLaughlin (1992) asserts that any nonlinear compositional principle is incompatible with Mechanism. Granting that was true for the historical figures discussed, it is not a condition that should be imposed as a general principle on generative atomism.

3.2 INTERNAL AND EXTERNAL RELATIONS

Much has been written about the British Emergentist movement, a great deal of it illuminating. Here, I shall explore an aspect that has not received as much attention: the role played in the early part of the twentieth century by the debate over internal and external relations within the nascent logical atomism movement.[12] This part of the historical background is important for understanding why the inheritors of the logical atomist tradition became resistant to the introduction of emergent properties and relations, and how, if we understand the limitations of that tradition, we can make sense of ontological emergence. In a later era, methodological individualists, famously antipathetic to the kinds of enforced social holism embodied in Marxism, also rejected internal relations and became another source of the opposition to emergence.

Logical atomism was one of the defining influences of twentieth-century philosophy. In its broadest form, the world was taken to be composed of atomic facts, which were either simple, noncomplex entities possessing a monadic property or a collection of simple, noncomplex entities related by an indecomposable relation. These atomic facts were then representable by atomic propositions having a structure that paralleled the structure of the atomic facts. All other propositions representing molecular facts were to be constructed from atomic propositions by the logical operations that are now familiar from basic logic courses. The true atomic propositions allow, in some versions of logical atomism, the derivation of all other true propositions.[13] Logical atomism was too simple to represent the world's structure and content, and its contemporary descendants are far more sophisticated. Yet many of them retain a commitment to two basic features of logical atomism that interfere with our understanding of emergence. These are the immutability of the basic elements and the treatment of relations. An additional legacy,

12. A discussion of the role played by external relations in what the author calls "Platonic Atomism" can be found in Hylton (1990, 113ff). Hylton does not discuss emergentism explicitly, although see p. 114.
13. Propositions were originally for Russell constituents of the world, although he later abandoned that view because of problems with false propositions.

the emphasis on synchronic relations, has led to difficulties that can be avoided by adopting a diachronic approach to interactions.

There were two aspects to the early twentieth-century debate over the status of relations that are relevant to our interests. The first was whether relational predicates could be fully reduced to combinations of monadic predicates. This question was decided negatively; there are essentially relational predicates, such as the arithmetical "greater than" relation on an infinite domain, that cannot be reduced to monadic predicates.[14]

The second debate, and the one of most interest to us, concerned the scopes of internal and external relations. I shall describe internal and external relations in more detail shortly, but for the moment we can take an internal relation in G. E. Moore's (1922, 291) sense, which is that an entity is changed in virtue of standing in an internal relation to other entities. In contrast, entities remain unchanged when they stand in an external relation to the other relata. This distinction became important around the turn of the twentieth century because idealists were promoting the concept of internal relations, partly with the aim of establishing a form of widespread holism, especially with regard to cognition.[15] If there were an internal relation between a cognizer and what was cognized, in the sense that each relatum would be different were it not in that internal relation, then the external world would be dependent for the way it is on its relation to the contents of our thoughts. This would lead to a characteristic feature of idealism, which is that reality is at least in part dependent upon human cognition.

14. An important difference between a first-order language containing only monadic predicates and a first-order language containing at least one binary relation is that the former is decidable whereas the latter is not. This difference stems from the differences between finite and infinite models, rather than anything connected to emergent features.

15. The figure most closely identified with this argument was F. H. Bradley, although there is disagreement about the extent to which he was committed to advancing the argument. Scholars of early twentieth-century philosophy tend toward the view that Moore's criticisms of Bradley missed the point, in that Bradley's original position was that all relations are unreal, rather than that all relations are internal (see Baldwin 1990, 33; and Hylton 1990, 54). I am interested here in the intrinsic merits of Moore's account of internal relations, rather than whether he interpreted Bradley correctly.

If one is interested in refuting idealism, then establishing the position that many relations are external was a sensible goal.[16] There is a related feature of internal relations in the social realm. If an individual stands in an internal relation to the society in which he lives, he will be changed as a result, and this change looks remarkably similar to an effect of downward causation if the society is at a higher level than the individual.

Moore was opposed to holistic aspects of reality long before his mature formulation of the position in his 1922 "Internal and External Relations" article. In his 1903 book *Principia Ethica,* when discussing things that have value, he notes that "the part of a valuable whole retains exactly the same value when it is, as when it is not, a part of that whole" (30). This anticipation of his later view that most relations are external is accompanied by the claim that "The value of a whole must not be assumed to be the same as the sum of the values of its parts" (28), and asserts that this is what should be meant by the Hegelian "organic whole." I note here the connection with the "whole is more than the sum of its parts" tradition of emergence. Moore denies that what is meant by an "organic whole" is that the properties of the parts "would not be what they are but for the existence of the whole" (33–34). He also argued that all relations were either logical or causal, and in both cases were external. This was more than a counterattack on Absolute Idealism; it was part of a long-term argument designed to keep the fundamental ontology free of holistic elements.[17]

Spatial relations are a core example of external relations, and they have the characteristic feature that a change in spatial relations alone

16. Hylton (1990, 121) writes "Moore follows the Idealists in giving relations and propositions the same status. He thus holds relations to be, like propositions, both real and non-mental. Second, all relations are external (I shall mention an exception to this later). Moore's dichotomy between the logical and the causal is directly connected with his opposition to internal relations. If all relations are either logical or causal, and neither logical nor causal relations are internal, then of course all relations are external." As David Sanford pointed out to me (personal communication), in Moore 1922 he is more cautious and argues only that some relations are external. The exception noted by Hylton is, interestingly, the relation of a complex whole to its parts (Hylton 1990, 143).
17. There is some dispute about just when Moore adopted his mature position on relations. See Baldwin (1990, chap. 1, sections 6, 7).

does not alter the intrinsic properties of the objects standing in those relations. If I move two sheets of paper farther apart, they remain the same sheets of paper with the same intrinsic properties. (One must think of objects isolated in space with only spatial relations holding between them. Where there are spatially dependent features of other kinds, such as electrostatic forces, this invariance may not hold because intrinsic properties of the relata might change as a result of those spatially dependent features.) Although what counts as an intrinsic property is disputed, I shall here take an intrinsic property to be a non-relational property—that is, a property the specification of which requires reference to no other properties or individuals. There are counter-examples to this identification, but they do not affect the arguments in this section.[18]

This debate over internal and external relations is still of interest because most contemporary metaphysicians, following the analytic tradition, reject internal relations in Moore's sense. In line with this antipathy to Moorean internal relations, the contemporary tradition of Humean supervenience has insisted that the relata remain unchanged whatever the type of relation involved. As a result, the nomenclature of internal and external relations has been changed so that almost all relations are now classified as internal under the revised definitions and are determined by the intrinsic properties of the relata. In what follows I shall qualify references to types of relations so as to be clear which sense of "internal relation" is at issue.

To subscribe to this rejection of Moorean internal relations, a great deal of reconstructive metaphysics must be brought into play, for the rejection conflicts with our everyday experience. Two people locked in a relationship of mutual hatred are psychologically different from how they were before they entered into that relation. Each player in a rugby scrum has his or her muscular properties altered in virtue of the mutual resistance of other players. Placing a blue circle on a red background

18. It would not do to take an intrinsic property instance as one that would persist were all other property instances in the universe removed, for then taking objects as bundles of property instances would be difficult. For example, a solitary property instance such as an electric charge rarely, if ever, exists without other, collocated properties, such as the spatial extension of the region where the charge occurs.

will result in the circle's visually vibrating, a property that the perceived circle would not have were it not in a proximal relation to the blue background. Parasites and hosts develop mutual dependencies, thereby altering their individual characteristics. A microphone and an amplified speaker, standing at a fixed distance, take on different characteristics when the power is turned on and a feedback relation begins. And the internal stresses in a flying buttress and a cathedral wall would be different were the two not in an equilibrium relation.

I have cited this wide variety of cases to show that Moorean internal relations are common when considered at the everyday level. Some of these examples involve dynamic relations, others involve static relations, and the changes in the respective relata appear, at least at first inspection, to be direct results of standing in the appropriate relation, rather than the relation itself resulting from the properties of the relata. So this sort of prima facie evidence in favor of active internal relations has to be explained away, and there are various ways that the generative atomist tradition can do this.

3.2.1 G. E. Moore

We can begin with G. E. Moore's famous treatment of internal and external relations in Moore (1922), although Moore was by no means the only philosopher who wanted to clarify the difference between the two types of relation.[19] For example, Russell rejected what he called the axiom of internal relations, which he expressed in uncharacteristically murky terms:

> "Every relation is grounded in the natures of the related terms.". . . If this axiom holds, the fact that two objects have a certain relation implies complexity in each of the two objects, i.e. it implies something in the "natures" of the two objects, in virtue of which they have the relation in question. According to the opposite view, which is the one I advocate, there are such facts as that one object

19. For simplicity here I shall talk in terms of two-place relations. Everything can be generalized to higher order relations.

> has a certain relation to another, and such facts cannot in general be reduced to, or inferred from, a fact about the one object only together with a fact about the other object only: they do not imply that the two objects have any complexity, or any *intrinsic* property distinguishing them from two objects which do not have the relation in question. (Russell 1910, 139–40)

Russell is here rejecting the existence of internal relations in a sense close to that discussed in Lewis (1986b) and restricting relations to those that are external in Lewis's sense.

Russell's denial that the intrinsic natures of the relata ground the relation between them manifests itself within the set-theoretical account of relations, which takes relations as sets of ordered n-tuples of objects. The intrinsic nature of these objects is irrelevant to whether the n-tuples can serve as a model of a given relation; natural numbers can serve as the relata in representations of relations that are initially taken to be about physical, biological, or other relations, for example. Standard model theory is thus inherently committed to doing without the axiom of internal relations that was criticized by Russell. As an example, consider a universe that contains three elementary particles, *a*, *b*, and *c*, where *a* is lighter than *b* and *b* is lighter than *c*. The ordered pair <*a,b*> is defined as the pair of unordered sets {{*a*}, {*a,b*}} and the relation "is lighter than" is in turn represented by the set {<*a,b*>, <*b,c*>, <*a,c*>}. This representation is purely extensional—that is, the relation consists of all pairs of particles for which the relation holds. Then, three things are important. The first is that although I have discussed this issue in terms of a representation, we could construe these relations realistically simply by allowing the existence of both sets and the particles that are the set members.[20] The second thing is that the ur-elements *a*, *b*, and *c* remain unchanged when they stand in the relation {<*a,b*>, <*b,c*>, <*a,c*>}, and in any other extensionally defined relation. Nothing about *a* and *b*

20. Allowing sets to exist in the physical world is problematical (see Chihara 1982) and allowing concrete entities such as particles to be members of abstract entities such as sets is doubly problematical. These are two reasons to doubt the extensionalist account of relations, but I am not advocating, only describing, the position here.

is altered in virtue of their standing in the "is lighter than" relation. This has repercussions for ontology because of the foundational importance of set theory. If the atoms of one's ontology are the ur-elements of one's impure set theory, then the foundations remain unchanged whatever relations are entered into. And the third thing is that the actual masses of *a*, *b*, and *c* are irrelevant to their ability to act as a representation of the "is lighter than" relation. That is, *c* could be lighter than *b*, which in turn is lighter than *a* and at the level of the model, we should still have a satisfactory representation of the "is lighter than" relation. This extensionalist account of relations, without further restriction, leads to all relations being external with nothing like Moore's conception of internal relations being even possible.

Moore's treatment of internal and external relations is notoriously obscure, but he was struggling to capture an important distinction; and it is unfortunate that his approach has all but been abandoned and replaced by the account of internal and external relations due to David Lewis. To keep these two treatments distinct, I shall use the terms "M-internal" and "M-external" to refer to Moore's concepts, and the terms "L-internal" and "L-external" to refer to Lewis's concepts.

This confusion surrounding internal and external relations has existed for a long time, and in a survey of these concepts by A. C. Ewing (1934), ten different definitions of an internal relation are listed.[21] I believe Ewing is right to count one sense as having priority. This is the view abandoned by Lewis but identified by Moore that an internal relation is one that "makes a difference" to its terms: "The assertion that all relations are internal in this sense means that, where the terms are related in some specific way, it is always true that they could not have been what they are without the relation being present" (Ewing 1934, 131).[22] The use of the modality "could not" in an era when modal

21. Not including, of course, more recent ones of the kind discussed below.
22. For an excellent treatment of some kinds of internal relations, see Dunn (1990), where in footnote 10 he remarks that Moore's definition allows two distinct readings, depending upon how we interpret "other than." The primary sense requires that for *x* to be other than **a**, *x* must have at least one intrinsic property that is different from those possessed by **a**. The secondary sense requires only that *x* be numerically distinct from **a**. That is, the primary sense differs from the secondary sense when the identity

semantics were improperly understood and treated informally is in large part responsible for the lack of clarity in Moore's position.

Moore's account of internal relations is no clearer in other passages: "Let P be a relational property and A a term to which it does in fact belong. I propose to define what is meant by saying that P is internal to A (in the sense we are now concerned with) that from the proposition that a thing has not got P, it 'follows' that it is other than A" (Moore 1922, 291). It is noteworthy that the modal claim is missing here. There is, then, a weak and a strong reading of the claim, the weak claim corresponding to modern definitions of an internal relation and the strong claim being radically different.

The weak reading is this: what is meant by an entity **a** being in an internal relation to **b** is simply that were the same relationship not to apply between **b** and a third entity **c**, then there has to be some intrinsic difference between **a** and **c**. Consider two positively charged objects **a** and **b**. Placed in proximity to one another, **a** stands in the relation to **b** of being repelled. There is a third object **c** that is spatially proximate to **b** but does not stand in the relation to **b** of being repelled. Then it follows that **c** is different from **a**. This is the weak reading of an internal relation, and it involves an application of the principle of the nonidentity of discernibles. The strong reading, which is supported by other remarks Moore made, asserts that it is in virtue of being in the internal relation that some of the intrinsic properties of an entity **a** are different. That is, either being related to another object, or the relation itself, produces a change in **a**.

Consider certain intrinsic psychological and physiological features possessed by humans who are in love with someone else—an elevated

of indiscernibles restricted to intrinsic properties is violated, in that for the secondary sense, *x* and **a** are not identical, yet share all intrinsic properties. Obviously, since *x* and **a** differ with respect to a relational property, the full identity of indiscernibles principle is not violated. Contrast Moore's definition with this passage from Goodman (1951, 40): "Two entities are sometimes said to be internally related if neither could be completely destroyed without affecting the other. . . . It seems to me to say that two individuals stand in such a relation is just to say that they have a common part. In other words, an *internal* predicate is one that applies between overlapping individuals only. An *external* predicate applies between discrete individuals only." (For other definitions, see Ewing 1934; Hylton 1990, 54–55.)

sense of happiness, a tendency to daydream, whatever collection of such intrinsic properties L you consider to characterize people who are in that relation to another person.[23] Now, suppose that **a** is in love with **c** and that **a** has properties L, but that **b** is not in love with anyone and lacks L. In this strong reading, it is because **a** is in the love relation to **c**—call this relation R—that **a** possesses L, and it is because **b** is not in the relation R that **b** is different (at least with respect to L) from **a**. Note that the love relation is not defined by the properties L. R is a binary relation that makes essential reference to, is directed toward, some other individual, whereas L consists of intrinsic properties only.

The strong interpretation comes very close to committing itself to the relation exerting a causal effect on one of its relata. Whether this is, in addition, an example of downward causation depends upon whether the relation is at a "higher" level than its relata. In some cases, the relation would be a member of the same domain as its relata—the relation of mutual repulsion entered into by a pair of similarly charged physical objects is a physical relation as are the objects themselves. In other cases, the relation seems to embed the related objects in another domain, as when humans enter into specifically social relations. Finally, it is a separate matter whether, in the strong reading, the changes in **a** persist when **a**'s relationship to **c** no longer holds, or whether **a** reverts to his or her former self. Although one could develop an interpretation of the strong reading in which the changes to **a** persist, the natural reading of Moore's own view is that they do not.

The strong reading of an M-internal relation leads to a specific kind of holism. If **a** would be (a) different (thing) were it not in the relation to **c**, then, assuming **a** exists, it is essentially related to **c** and so is necessarily part of the related pair <**a,c**>. This entails that it is impossible for **a** to exist in isolation and so deducing how it will behave in the relation from how it behaves in isolation is moot. This strong interpretation proved unpalatable to many philosophers and pushed them, as it did Moore, away from the view that M-internal relations exist and, later, to firmly resist the use of holistic emergent properties in our ontologies.

23. Self-love could be an internal relation under the strong reading, but that relation potentially involves self-referential complexities that are best avoided here.

Finally, I note that if the atoms are of only one type, as in Broad's Pure Mechanism, there could be at most one M-internal relation type instantiated at the atomic level. The existence of two distinct M-internal relations would require the relata of each to be different within the two relations, which is impossible. Moreover, in such a world, the existence of a single atom that is not in an M-internal relation would preclude the existence of any M-internal relations.[24]

3.2.2 Lewis's Definitions

An important part of David Lewis's doctrine of Humean supervenience is the rejection of emergent properties: "there might be extra, irreducible external relations besides the spatiotemporal ones; there might be emergent natural properties of more than point-sized things. . . . But if there is suchlike rubbish, say I, then there would have to be extra natural properties or relations that are altogether alien to this world" (1986a, x).[25]

What Lewis means here by an external relation is different from what Moore meant. Here are Lewis's definitions of the two kinds of relations:

> An *internal* relation is one that supervenes on the intrinsic nature of its relata: if X_1 and Y_1 stand in the relation but X_2 and Y_2 do not, then there must be a difference in intrinsic nature either between the Xs or else between the Ys. If X_1 and X_2 are duplicates (or identical) and so are Y_1 and Y_2, then the pairs $<X_1,Y_1>$ and $<X_2,Y_2>$

24. I assume here that if relations R_1 and R_2 have relata that in all instances are of the same type, one must be a subrelation of the other, differing only in how many pairs each applies to. It is possible to avoid this either by using some intensional characterization of relations or by appealing to counterfactual instantiations, but I shall not pursue that here.
25. Lewis allowed that there were non-spatiotemporal external relations, one example being the relation of nonidentity (1986b, 77), although that example has the defect of not being a natural relation. He also recognizes that there might be physical counterexamples to his view, but he dismisses them with the remark: "But I am not ready to take lessons in ontology from quantum physics as it now is. First I must see how it looks when it is purified of instrumentalist frivolity . . . and when it is purified of double thinking deviant logic and . . . when it is purified of supernatural tales about the power of the observant mind to make things jump" (1986a, xi).

> stand in exactly the same internal relations. . . . I shall say that a relation is *external* iff it does not supervene on the natures of the relata taken separately, but it does supervene on the nature of the composite of the relata taken together. (1986b, 62)[26]

It has been noted that this definition is not fully satisfactory until we are told what "the nature of the composite of the relata taken together" means without circularity. Some examples of applying these definitions are the relation of intrinsic similarity, which is L-internal; the relation of spatial distance, which is L-external; and the relation of having the same owner, which is neither L-internal nor L-external. The fact that external relations in Lewis's sense do not supervene on the intrinsic properties of the relata alone might be a reason under some criteria of autonomy to consider L-external relations such as spatial relations candidates for emergence. Some philosophers have in fact argued that spacetime structure emerged during the initial expansion of the universe, but I shall not pursue that here.

The relation between Lewis's definition of an internal relation and Moore's is straightforward. Suppose **a** and **c** are in an L-internal relation R. Let the supervenience base for R consist of the intrinsic properties $F_1, \ldots, F_n$ for **a** and $G_1, \ldots, G_n$ for **c.** For the relation R between **a** and **c** to supervene upon these properties means that the presence of $F_1, \ldots,$ F_n and $G_1, \ldots, G_n$ necessitates the presence of the relation R between **a** and **c.** Hence, if **b** were not related by R to **c**, then **b** must lack at least one F_i, given that **c** has $G_1, \ldots, G_n$ and, via equation [3.1′] of section 3.2.4 appendix), it would not be **a.** And this gives us that **a** is M-internally related to **c** in the weak sense. It is implausible that it is M-internally related to **c** in the strong sense because the direction of supervenience determination is from the relata to the relations.

26. Some associated definitions are required to understand the above account: "two things are *duplicates* iff (1) they have exactly the same perfectly natural properties, and (2) their parts can be put into correspondence in such a way that corresponding parts have exactly the same perfectly natural properties, and stand in exactly the same perfectly natural relations. Then we can go on to say that an *intrinsic* property is one that can never differ between two duplicates" (Lewis 1986a, 61–62).

The converse entailment between an M-internal relation and an L-internal relation does not hold. Suppose **a** and **c** are M-internally related by R, but **b** and **c** are not. There must be at least one difference between **a** and **b**, but it need not be a difference in an intrinsic property. Let **a** be a person who is married to **c** and who is such that **a** would be different if **a** were not married to **c**. Take a third individual **b** who has the same intrinsic properties as **a** (i.e., **a** as **a** is within the marriage relation) but who is not married to **c**. So the "is married to" relation is not L-internal. But that relation is M-internal, since **a** would be different if not in the relation.

Because Lewis's definition requires that the L-internal relations supervene on the intrinsic properties of the relata, if we use as relata entities that have too few properties, a given instance of a relation may not be L-internal, but may become so when properties are added. Dissimilarity, for example, would not hold between entities with the same size and shape, but would hold when the property of redness was added to one but not to the other. However, we can adopt the Leibnizian principle that any concrete object possesses a maximally consistent set of natural properties, positive and negative; and so this problem with missing properties will only appear in our claims about L-internal relations, not in the relations themselves.

Are spatiotemporal relations and L-internal relations sufficient to account for all features, functional and otherwise, of our world? Consider a complete set of parts that need assembly to function as a chair. The chair comes with an epoxy glue that requires two liquid components to be mixed, which then harden by a chemical reaction to bond the components of the chair. It is not enough for the parts to stand in the right spatial relationships to one another, for I could carefully place all the pieces of the chair together, using glue in the liquid state so that, left untouched, the composite has all the right appearances of being a chair. But any attempt to use it as a chair would result in disaster. Once the glue has cured, the same components of glue and wood, standing in exactly the same spatial relations, now exhibit a novel property—the property of being a functioning chair, together with the property of rigidity for the whole.

There is what seems to be an obvious answer to this example, which is that the glue has different properties when cured than it does when wet, that these properties are local and intrinsic, and that the relation "*x* is glued to *y*" supervenes upon these local intrinsic properties. Yet this response is not so obviously correct as it might seem. For what are those local intrinsic properties of the glue? Not simply being hard (or dry) as opposed to liquid (or wet), for merely placing dry glue in the same spatial relations as the cured glue will not give us the right load-bearing properties needed for the components to function as a chair. You must go down to the microscopic level to capture the bonding relation between the glue and the wood, and here is where troubles begin for the anti-emergentist. This is because the epoxy reaction forms covalent bonds between the resin and the hardener, which as we have seen exhibit ontologically emergent features. These are not amenable to either an L-internal or an M-internal relational treatment, as the nature of covalent bonding is not one that has a relational form.

The example I have just described is far from unique. Ordinary experience provides us with many examples of apparently emergent phenomena that proponents of generative atomism claim can be shown not to be emergent on a finer analysis. But when the anti-emergentist resorts to a finer grained microscopic description of the example to show that the property in question is not after all emergent, and ordinary experience is discounted in favor of scientific theory, the anti-emergentist has to assume that generative atomism will reappear as a viable approach at the microscopic level. And that is exactly what does not always happen.[27]

Now compare the definition of an L-external relation with a criterion for being an external relation that seems to differ from Lewis's earlier definition: "What makes a relation external, I would have thought, exactly is that it holds independently of the intrinsic natures of the two relata" (1986b, 180). Whether this definition is coextensive with the first definition of an L-external relation depends upon whether the phrase

27. The Humean supervenience approach is compatible with the elements of the supervenience base's being different when the relation holds than when it does not. Thus, it can be true that R holds whenever **a***, **b*** occur in context C (where C is characterized other than through R), even though **a** and **b** change to **a*** and **b***, respectively, when entering C. The fundamental entities arrange themselves into the appropriate structure and the L-internal relation then comes for free.

"does not supervene on the natures of the relata taken separately, but it does supervene on the nature of the composite of the relata taken together" denies or allows that the intrinsic natures of the relata are taken in addition to the nature of the composite. For consistency, I shall interpret the first definition in the former sense.

Almost all relations are M-external for Moore, but under Lewis's classification very few relations are L-external. So why is there this difference between Moore and Lewis on how to classify most relations? The answer is this: generative atomists require not just a conceptual but also an ontological priority for their atoms. First fix the atomistic base, then all the rest, including the relations, will be fixed, too. The supervenience formulation does this when we have an L-internal relation. Moore wanted to avoid cases in which the relation itself affects atomistic relata. And the only way to do that is to argue that the atoms are never affected by being in the relation, so there are no M-internal relations on the strong reading, and hence all instances of relations are such that the relata are not altered in virtue of being in the relation. As such, the relations are all candidates for being either L-internal or L-external relations, and as it happens, almost all of them are L-internal. To put it slightly metaphorically, for Moore, the direction of influence is never from the relation to the relata.

3.2.3 Relationalism, Holism, and Interactions

In section 3.2 we caught a glimpse of the connection between internal relations and the holistic approach to emergence through Moore's denial of the sense of "an organic whole" in which the properties of the parts "would not be what they are but for the existence of the whole." There are echoes here of holism, but the holistic terminology is different from the interactive approach used in transformational emergence. This is because in the latter it is, in the nonspontaneous cases, the effect of interacting with at least one other thing that is important, not being part of the whole. Conversely, it is possible for something to be changed merely as a result of being part of a whole without interacting with the other parts of that whole.

For example, take a legislative body that must have a quorum of six members present in the chamber and simultaneously voting for there to be a binding legal decision. At time t, there are five members of the body in the chamber, not including Emily. Emily enters the chamber at t', does not speak or otherwise interact with the other five members, and all vote in favor of the motion at t'' when the bell sounds. In virtue of being part of the whole, Emily has taken on a property that she lacked at t—someone who helped pass the legislation—and she would not have had that property were she not part of the whole group of six.

Finally, recall Teller's formulation of relational holism discussed in section 2.3.3: "By *relational holism* I will mean the claim that objects which in at least some circumstances we can identify as separate individuals have *inherent relations*, that is, relations which do not supervene on the non-relational properties of the two individuals' (1986, 73). The examples described in Teller's article can be taken as additional examples of L-external relations.[28]

3.2.4 Appendix

Following Baldwin (1990), we can construe an essentialist view of internal relations as

$$xRy \supset \Box(u = x \supset uRy) \tag{3.1}$$

That is, if x and y stand in an internal relation in our world, then in any possible world in which x and y exist, x stands in the same relation

28. Teller himself does not put the point this way nor, because his article was published in the same year as Lewis's (1986b), does he mention Lewis's position. Teller does independently state a position that is very close to Lewis's formulation of Humean supervenience (which is articulated in Lewis 1986a), and identifies it with what he calls classical physicalism: "I suggested earlier that we understand classical physicalism as local physicalism, that is as the supervenience of all the facts on the non-relational physical facts about physical objects" (1986, 80). Teller cites an unpublished paper by Allan Stairs, and published papers by Shimony (1978), and d'Espagnat (1973), as containing prior discussions of relational holism (76n1).

to y.[29] It perhaps better captures the intent of Moore's idea to use the formulation:

$$xRy \supset \Box\left(\neg\, uRy \supset \neg u = x\right) \qquad (3.1')$$

where I assume that contraposition is valid within the modal operator. Note that in order to capture the contingent nature of some internal relations, the necessity will in some cases be natural necessity rather than metaphysical necessity.

Employing the discernibility of non-identicals applied to intrinsic properties only, we have, generalizing to both relata:

$$\mathrm{xRy} \supset \Box(\neg \mathrm{uRv} \supset [\exists \mathrm{S}(\mathrm{Sx} \,\&\, \neg \mathrm{Su}) \mathrm{v} \exists \mathrm{Q}(\mathrm{Qy} \,\&\, \neg \mathrm{Qv})]) \qquad (3.2)$$

where S and Q are intrinsic properties.

Equation (3.1) is to be contrasted with Leibniz's law:

$$\Box\left(xRy \supset \left[u = x \supset uRy\right]\right) \qquad (3.3)$$

Equation (3.1) requires that any contingent relation holding between two relata in our world—say, a causal relation—will hold in any other world where the relata exist. In contrast, equation (3.3) only requires that where a relation holds within a world, then that relation must be preserved under any substitution of identical objects within that world. The weak reading of an M-internal relation is exactly captured by equation (3.1′).

Some care needs to be taken to prevent the claim from becoming a triviality. If we count any property, monadic or relational, as playing a role in the identity conditions for an object, then for any relation, the mere fact of not being in that relation will be sufficient to change the identity of the former relatum. In the case of Cambridge changes, an example of which occurs when I lose my status as an uncle in virtue of

29. Moore's own presentations are ambiguous about whether an internal relation must alter all of its relata or just some. It is best to adopt the weaker account in which at least one relatum is altered.

my sole nephew's or niece's dying, whether or not I am changed in virtue of losing that relational attribute is a matter for legitimate dispute.[30]

An intimate connection with essential properties is maintained by some analysts. We have, for example: "Roughly . . . a relation is internal to an object if it is part of that object's nature or essence to be related in that way; if it is accidental or not a part of the object's inner nature to be so related then the relation is external" (Hylton 1990, 54); and "An internal relation is one that is in some way essential to, or constitutive of, the identity of the objects related, so that without the relation neither would be that object which it is" (Hylton 1990, 121)[31] This account risks making internal relations description relative; the relation "being an uncle of" is not internal on a de re interpretation of "Paul Humphreys" (it is, in a robust sense, accidental that I am an uncle), but not on a de dicto reading under which Uncle Paul cannot but be related to his only nephew. The key thing is whether the relatum is the same within the relation as outside it. There is much more to be said on this, but whatever the reading taken, this account is different from a contextualist approach, and here I merely note that logical treatments of this and similar issues often fail to distinguish conceptual issues from ontological issues. Conceptually, certainly in terms of the definition of "is an uncle," I have lost the relational property even if I remain completely unchanged in terms of monadic properties. Causally, the loss of my nephew could result in my mourning the deceased, or having to move out of the house that my late nephew had allowed me to use, and so on.

Concerning equation (3.2) above, in the non-essentialist version, the change consists in a change to at least one intrinsic property of the individual. In the essentialist version, in which we separately identify the essential properties of an entity, the change consists in the loss of an

30. To be clear, "is an uncle" is at least a binary relation, and not a monadic property. (I say "at least" because one may wish to add a third relatum, consisting of a system of legal principles, because whether one retains the legal status of being an uncle after the death of the last remaining nephew or niece may vary depending on the legal code governing the situation. For the biological relation, the considerations are different.) One definition is: "x is an uncle if and only if x is a male and x has a sibling y such that y has a child, or x is married to someone who has a sibling who has a child."

31. See also Rorty (1967).

intrinsic property that is essential to that object qua type. The essentialist version captures at least one of our previous examples. In extreme cases of mob behavior, we do think of the members of the mob as being essentially different in virtue of entering into various relations and undergoing certain interactions that result from being embedded in the mob. Once outside the mob, and no longer in those relations or interactions, they lose the property of being violent and regain their identities as humans. In contrast, the case of fusion, with the going out of existence of the entity, cannot be represented by equation (3.2) because *u*, *v* lack referents.

Is it possible for two entities to be internally related in one world but not in another? The intent of the essentialist interpretation is that this is not possible. Take the relation that any helium atom is intermediate in atomic weight between hydrogen atoms and lithium atoms. This is an internal relation, in that anything that was not intermediate in atomic weight between hydrogen and lithium not only would not be helium but also necessarily would not be so. This is a result of our conceptual classification of the elements, if atomic weight is an essential part of that classification. In contrast, this condition is too strong for the non-essentialist reading. The two psychological adversaries of one of our earlier examples, however strongly they may feel, need not be in a relationship of mutual hatred. The relations themselves are not invariantly internal in the sense that a relation may be internal to one set of objects yet not to another: lying between the United States and Canada is a relation that is internal to the United States–Canadian border, but this is not a relation that is internal to me when I am crossing it.

3.3 LEVELS

The concept of levels is so embedded in discussions of reduction and emergence that it is tempting to take such talk literally. Levels are what make the ideas of downward causation and upward causation seem natural, and the literature makes heavy use of diagrams containing horizontal strata, downward-pointing arrows, and vertical determination relations. One could come away with the view that downward causation is analogous to calling down to the basement to let your spouse know that dinner is ready. There is some merit to the metaphor of levels, but it is beset by difficulties, some of which I describe below.

The levels concept is hard to avoid even within the diachronic accounts. Theories of evolutionary emergence appealed to different levels of ontology, and life was at one time considered to be emergent in both the temporal and the hierarchical senses, having emerged at some point in the development of the initially purely physical world. Similar considerations held for mental phenomena of various kinds. The central feature of these earlier positions on emergence was that there are at most half a dozen levels (physical, biological, psychological, social, perhaps spiritual or other levels) and each new level was radically different from its predecessors. Recent developments in physics strongly suggest that even within the layer of the physical, there are multiple sublevels, each emerging in turn from interactions between instances of properties at lower levels. This increase in the number of levels is consistent with the use of effective field theories (EFTs) that has given rise to a nested series of models within particle physics.

To avoid talk of levels, I have adopted in this book a framework of domains rather than levels. The domains approach has the important advantage that it allows us to initially remain neutral about how the entities of two domains are related. The levels imagery is shot through with synchronic concepts, most obviously compositionality, and there is a prior ontological or methodological preference given to the lower levels over the higher, an asymmetry that is tacitly employed in some arguments for physicalism. The mereological account works reasonably well for some parts of physical ontology, but even if we accept this limitation, the mereological hierarchy breaks down once we reach the realm of the psychological, because there are no obvious part/whole relations that hold between the ontology at the neural level and characteristically mental entities at the psychological level, such as affects or propositional attitudes in cognitive psychology. The inclusion hierarchy picks up again with the social sciences if we consider social agents to have psychological properties, but there is no clear ontological hierarchy involved in the relations between economics, sociology, politics, and anthropology, and so we at best have a partial ordering of levels.

In contrast, the domains approach allows us to speak of the elements of one domain being reduced to the elements of another, or the elements of one domain emerging from elements of another, without being locked into a total or partial ordering of domains. This allows room for diachronic approaches to emergence, and it has the additional advantage of forcing us to explain what we mean by "downward causation" and "upward causation." Relationships between the domains will usually exist, because certain phenomena of biology, such as the presence of the protein mHtt, allow the prediction of some of the cognitive effects of Huntington's chorea—effects that would usually be considered to occur in a different domain from those to which their causes belong, but these relations will often be piecemeal. We thus need to decide, if we want to talk of levels, whether there is an effective criterion that can be used to separate levels, and whether this talk of levels is to be taken as accurately representing real divisions in the world, or if it is merely a metaphor. It might be that only some of the levels objectively exist, the others being merely convenient ways to represent parts of the world.

The most common criterion for levels in the literature on synchronic reduction and emergence is based on a simple compositionality approach.[32] Physical atoms are supposed to contain elementary particles as constituents, molecules are supposed to contain atoms, macromolecules are made up of smaller molecules, biological entities are built from macromolecules, and so on. This approach has the advantage that if a criterion for levels can be given in terms of objects, then a hierarchy of properties can be tied to object levels in the following way: if, in all levels below j, no instances of property P occur, but there is at least one instance of P at level j, then P has level j. This approach requires taking objects as ontologically more basic than properties. Attributing levels to laws is more problematic. The criterion for properties could be used

32. "What gives this array structure is the mereological relation of *being part of*: entities belonging to a given layer are mereologically composed of entities belonging to the lower levels, and this relation generates a hierarchical ordering of the levels" (Kim 1993b, 337).

for attributing levels to laws in the following way: if the law L involves all and only properties $P_1, \ldots, P_n$ at levels $l_1, \ldots, l_n$, respectively, then L is assigned level $\max\{l_1, \ldots, l_n\}$. Yet some laws, such as psychophysical laws, ought to have more than one level associated with them—those that they bridge. Perhaps such laws that connect levels should be placed in a separate category, but so many laws concerning properties of one level make use of properties at other levels that this would restrict the levels position to laws of "pure biology" and so on.

Setting aside for the moment the deficiencies of compositional ontologies—molecules are not composed of atoms in anything like the way the solar system might be thought to be composed of the sun, planets, moons, asteroids, and so on—the compositional inclusion of A in B cannot be a sufficient condition for A to occupy a lower level than B because it would produce far more levels than are usually considered to be present in nature. The addition of one stone to a heap would put the new heap at a higher level than the old. As a sufficient condition, it also leads to arbitrariness. Take a composite object made up of two stones. This object will be at a higher level than either of its constituents. Consider breaking one of the stones that constitutes the two-stone aggregate into three pieces, but let two of the fragments remain fitted together. Do we now have three constituents of the original group or four? Depending on the answer, the original pair of stones has been moved up either one or two levels.

Moreover, what it is for A to have B as a constituent is a subtle matter. Mere spatial inclusion of A within B is insufficient for constituency, since my sitting in the midst of the group comprising the Supreme Court does not make me part of that legal body. The example involves functional roles, but many nonfunctional entities employ such criteria. A soccer ball kicked into a cave is not part of the cave, and even within the functional realm, an intravenous catheter inserted into my forearm is not part of my body, even though spatially and functionally it is wholly within it.

Can we use compositional spatial inclusion as a necessary condition for one object to be a constituent of another? Unfortunately, this will not work, either. The passenger door of my car is currently spatially separated from the rest of the car, being in the body shop, yet it is still part of my car. Widening the spatial region to include the body shop would mean that many other things in the region, such as the engine from a Ferrari, are spatially included, but it is not, alas, part of my car. Finally, spatial inclusion is not a well-defined notion for many physical systems such as electrons.

Perhaps we could weaken the criterion by dropping the spatial inclusion condition: O′ occupies a higher level than object O only if object O′ has object O as a constituent. One could impose this condition, but this would require giving up the position that emergence always involves a change in levels. Consider the phase transitions vapor → water → ice, which are commonly cited as examples of emergence. The mereological approach to levels will place a collection of H_2O molecules having the property of solidity and the same collection of H_2O molecules having the property of liquidity at the same level. Since the phase transitions are invertible, either can emerge from the other, showing that emergence does not always involve a change in levels.

Mereological inclusion is, thus, not a good candidate for identifying levels. One could appeal to the presence of new laws. By adding the requirement that the higher level entities obey different laws than do the lower level entities, the hope is that the compositionality criterion can be converted into a necessary and sufficient condition for identifying levels. As a candidate definition, we have that entities of type B occupy a higher level than entities of type A if and only if the entities of type B have type A entities as constituents, there is at least one law or theory L that applies to type B entities that does not apply to type A entities, and L cannot be reduced to the laws applying to type A entities. Here, some account of the ontological reducibility of one law to another would need to be provided, because the derivability of one law statement from another is too weak a condition—mathematical difficulties could prevent the derivation even though the underlying laws stand in a semantic dependency relationship. This is perhaps the most promising account of levels available, but it still depends upon the availability of the compositionality criterion. Without that criterion, the attempt fails because two types of simple fundamental entities, such as fermion and bosons, can be governed by different laws, neither of which is reducible to the other nor is there an even more basic unifying law, but the entities in question would not count as occupying different levels.

Energy—or, equivalently but inversely, length—has been used by effective field theories in physics as a basis for making an objective division of the physical world into levels, where an entity occupies a higher level only if it has a lower energy.[33] An important consequence of this approach is that by restricting the representation at a given length scale to specifically chosen degrees of freedom, the details of the levels below can be ignored. One way to do this is by using parameters such as density and viscosity that aggregate shorter distance behavior and make it possible to ignore the details at the micro level. A disadvantage to this method is that it makes the approach model dependent. Effective field theories are also only approximate theories because of the ignored degrees of freedom at those lower levels.

Finally, we have as a proposed sufficient condition that, if the existence or properties of an entity A can be explained in terms of another entity or collection of entities B, but not vice versa, then B has ontological priority over A. It is not a necessary condition because supervenience relations often do not permit explanations, but it is generally taken that if A supervenes on B, but not vice versa, then B has ontological priority.[34]

It is worth noting that Boolean combinations of properties never change the level which the properties occupy unless one holds the view that there is a threshold of complexity beyond which emergence occurs. It might be objected that a related principle does not hold for set theoretic operations, and set formation is a basis for the standard set-theoretical hierarchy. But that set-theoretic hierarchy is not the one that people have in mind when they talk about levels of properties. Biological properties are supposed to occupy a level above physical properties, yet as set-theoretical entities, both physical and biological properties are at the same level unless one construes atoms as sets of elementary particles, molecules as sets of sets of elementary particles, and so on. It also bears noting that the ordering of property levels does not coincide with the ordering of the sciences. It was common at one time to rank the sciences in a strict hierarchy and many people have their own rankings,

33. See Hartmann (2001) and Castellani (2002).
34. This principle is endorsed by Klee (1984).

usually with physics as the most fundamental.[35] However, most sciences use a mixture of properties from other sciences; for example, economics ought to include factors from at least climatology, sociology, anthropology, physics, and chemistry in its models to effectively predict commodity futures on a fundamentalist approach, and multiscale modeling is common in areas such as climate modeling.[36]

In light of the above discussion, I believe we are better off rejecting the existence of clearly defined levels and using instead the more neutral device of domains.

3.4 DOWNWARD CAUSATION

Donald Campbell is usually credited with introducing the idea of downward causation. Campbell originally applied the idea to issues involving natural selection: "Where natural selection operates through life and death at a higher level of organization, the laws of the higher-level selective system determine in part the distribution of lower-level events and substances . . . all processes at the lower levels of a hierarchy are restrained by and act in conformity to the laws of the higher levels" (1974, 180). The principal example given by Campbell is the role played by social processes in the evolutionary success of termite groups, whereby worker termites feed the soldier termites, the jaws of which are so specialized for attack purposes that they cannot feed themselves. This example is not wholly compelling, because unless we are willing to attribute collective intentionality to groups of termites, the division of labor that underlies the supposed separation into levels is simply a physical differentiation in the behavior of two different biological populations of individuals that results from natural selection.[37]

35. See, e.g., Whewell (1840) and Comte (1830). For a philosophical assessment, see Humphreys (1990).

36. A criterion that is not intended to be either a necessary or a sufficient condition is: a characteristic of a level is that it (a) exhibits stable features, and (b) has novel causal powers. One approach along these lines can be found in Wimsatt (1994, 210–11).

37. For discussions of collective intentionality, see Tuomela (1995) and Searle (1995). Campbell's arguments relied on compressing time periods that involved several

Can downward causation exist, even though this particular example is not completely persuasive? Two problems about downward causation make its postulation problematical, each of which is considered serious enough to preclude the existence of this type of causation.[38] The first source of worry is that if a given system can causally affect its own components, via holistic properties that have downward effects, a causal loop will be generated that violates the irreflexive property of causation because the cause includes the effect as one of its parts.[39] This worry about whole/part causation leading to partial self-causation is an artifact of the emphasis on synchronic emergence, for the heart of the objection is that downward causation would involve a simultaneously acting causal loop between the whole and at least one of its constituents.

Downward causation is not obviously problematic when diachronic processes are involved, because a causal process that originates with the whole system can affect one of its components at some later time without violating the irreflexivity condition. If, in my capacity as a voting member of a committee, I help pass (by a single vote) a policy that later results in my being dismissed from the committee, I was a cause of my own firing via the higher level committee; but there is nothing conceptually incoherent about this even if, for the sake of the argument, we grant that downward causation from the committee to the individual members exists in this case. In this example, an argument might be made for simultaneous causation—I vote last, the policy takes effect at the instant the last vote is recorded and has immediate effect—but this would result in my both being a member of the committee and not being a member of the committee at the same instant.

generations into a single instant, and so the causation claim is a model-level claim and not one that is directly about the world.

38. See Kim (1999) and Ladyman et al. (2007) for opposition to strong emergence on the basis of problems associated with downwards causation. Mark Bedau also rejects strong emergence on completeness grounds: "There is no evidence that strong emergence plays any role in contemporary science. . . . Strong emergence starts where scientific explanation ends" (Bedau 2002, 11).

39. See, e.g., Emmeche, Koppe, and Sjternfelt (2000, 14). Thomson and Varela (2001, 420), Silberstein and McGeever (1999, 182), and Yates (2009) also construe downward causation as involving causal influence of a whole on its parts.

We can capture this type of whole/part diachronic causation via the definition:

> *Definition* Let S be a system having components. Then an occurrence D(S,t) of a property D causes an occurrence E(y,t′) of a property E via whole/part causation just in case D(S,t) causes E(y,t′), *y* is a proper part of S, t′ is later than t, and D is not a property of any proper part of S.

The last condition is to ensure that the property D is a "higher level property." In the committee example, no member of the committee has the properties possessed by a committee policy on membership.

The second source of worry about downward causation occurs when the causation is from some "higher level" entity, but the effect does not involve a component of the system to which the cause is attached. One such example, assuming that the biological domain is at a higher level than the physical domain, is when the roots of the oak tree in my yard crack the brick foundations of my house by means of a growth process that is biological in origin. The problem with this kind of causation from domain A to domain B is not downward causation as such, but a conflict with the assumption that the domain B is causally closed. I showed in section 1.6 that causal completeness does not preclude domain-to-domain causation and that there are situations in which such causation can occur. So, we can conclude that postulating the existence of diachronic analogs of downward causation is not subject to the primary problems associated with synchronic versions of that concept.

3.4.1 Cube World

Another way to approach downward causation is to consider it as a set of constraints that do not operate at the individual level, but that are consistent with the lower level laws.[40] We can get a grip on what is involved by considering a simple ontology in which the atoms are cubes

40. E.g., Campbell (1990, 4).

with sides of length 1 inch. I offer this example not as a definitive case of downward causation, but as an illustration of the problems that must be addressed before a clear assessment of downward causation can be given in general.

The laws of this world permit the cubes to weld together to form solid cubes with sides of length N inches, where N is a positive integer, and also to form cubical lattices having sides of length 2N + 1 inches with apertures of size 1 inch. The lattice is rigid, the material of the cubes is impenetrable, and there is no friction. Larger cubes and lattices are composed of atomic cubes of size 1 inch. Can we make a case that the lattices and cubes occupy different levels or domains? Mere differences in size are not considered sufficient for a change in levels and cubes of size 1 inch and of size 2 inches, which both satisfy the laws for solid cubes, plausibly exist within the same level or domain. The lattices are different, however, because they obey a different law than do the solid cubes; the lattices allow some objects to pass through them, which the solid cubes do not. So, we could use this fact that a different law applies to lattices than to cubes to attribute a higher level to the former than to the latter. When the cubes are allowed to fall perpendicular to the lattice, only 1-inch cubes are found on the underside of the lattice. This is a simple example of an object, a lattice, existing at a higher, non-atomic, level placing a constraint on the behavior of objects at a lower level in a way that is perfectly transparent and unproblematic. The higher level property of rigidity plays an essential role in the explanation of why only 1-inch cubes are present below the grid.[41] If the grid were not rigid, it could deform and either larger cubes could get through or 1-inch cubes could be blocked. In this example, the constraints are causes of the effect because the counterfactual claim, "If the grid had not been rigid, the effect of only atomic cubes passing through would not have occurred" is true. So, this type of constraint by a higher level domain also consistently allows "downward causation."

41. Those familiar with Hilary Putnam's (1975) example of explaining why a square peg does not go through a round hole will see that an explanatory appeal to a higher level of theoretical description is present in that case. The example here is ontological and does not support the kind of pragmatic theory of explanation that Putnam advocated at one point in his career.

Are there downward constraints that are not also downward causes? A number of potential examples exist. Heylighen 2001 notes that in processes such as Bénard rolls, individual molecules are constrained to move in a given direction by the entire roll, and that as the Weiss domains form in ferromagnets, the orientation of individual spins is influenced by nonlocal fields generated by the entire domain. Laughlin (2005, 66) asserts that the motion of electrons and holes through semiconducting silicon is the collective motion of all the electrons that behave as if they were isolated, and that there is a detectable difference in the acceleration mass and the strength of the electric forces. "Metallic behavior is an emergent organizational phenomenon. The electron sea makes sense because the metallic phase has formed, not the other way around" (85).

Pauli's Exclusion Principle (PEP) is an example of a constraint that is holistic in form. PEP is responsible for the stability of matter because it prevents matter from collapsing into much smaller regions of space. It also forms the basis for explanations of why two macroscopic objects cannot be in the same place at the same time, and of the familiar periodic table of the elements.[42] When acting as a constraint on electron states, PEP has a holistic aspect, because a new electron added to an atom is constrained to be in a state that depends upon which states all of the other electrons are in. PEP is an essentially non-individualist law or constraint; it makes no sense to apply it to entities without reference to other entities, but it is not a relation with fixed arity, because the principle applies to an arbitrary number of fermions. Finally, PEP acts via constraints rather than using "downward causation" because it is a fundamental law.

3.5 HOLISM, CONTEXTUALISM, AND TRANSFORMATION

The quotation from Broad, which I cited in section 1.7.1, is characteristic of many positions on synchronic emergence. It suggests that emergence occurs when a relation imposes synchronic structure on a collection of

42. For an historical discussion of the origins of the PEP, see Massimi (2005).

entities and novelty appears as a consequence, whether ontological or inferential. I shall argue that this synchronic relational approach, and related conceptual apparatuses involving holism and contextualism, is ill-suited to the task of understanding emergence because it directs our attention away from important diachronic features.

Let us return to our definition of a nomologically holistic property:

> *Definition* P is a nomologically holistic property of a system S just in case P is possessed by S as a whole and as a matter of nomological but not metaphysical necessity, P could not be possessed by any of the proper parts of S.[43]

A weaker, contingent version requiring only that P is actually possessed by the whole but not the parts could be formulated, but it would fail to capture what is usually considered needed for emergence. Take a rock band that plays at a lower than usual sound level of 100 dB. None of its members is individually producing that sound level, but as a matter of nomological possibility, each could do so by turning up his amplifier to 11. The actual collective sound level is accidentally a property of the band as a whole and not of any of its parts, and therefore is not holistic in a sense relevant to emergence. In addition, the lack of novelty of the collective property of loudness counts against this as an example of emergence, even though the decibel level from *n* sources is a nonlinear function of the decibel level of the individual sources. Nonlinearity by itself is not a basis for attributing emergence to the system, since it is insufficient.

C. D. Broad did canvass the view, although he did not endorse it, that holism is sufficient for emergence.[44] It is easy to show that this is incorrect for our condition of nomological holism. Consider a chemical balance located in a uniform gravitational field with a 7-ounce weight on the left-hand pan and two 5-ounce weights on the right-hand pan.

43. Stephan (1999) calls such properties "systemic."

44. "[A]n emergent quality is roughly a quality which belongs to a complex as a whole and not to its parts" (Broad 1925, 23). Most writers in this period were less concerned with modal features than are modern authors.

The whole, consisting of the two 5-ounce weights, has the property of causing the right-hand pan to move downward, a property that nomologically neither of the 5-ounce weights individually can have. But the metaphysical possibility of this happening (in worlds in which there are different laws relating mass to weight) exists and hence the nomological necessity is not an instance of metaphysical necessity. So, this is an example of a nomologically holistic property, but it is not an example of emergence.

A related but stronger contingent aspect of holism applies to the well-known type of epistemological holism: evidential holism. An immediate consequence of the Duhem-Quine hypothesis, evidential holism is the claim that it is not individual sentences that are confirmed and disconfirmed by evidence but entire networks of theoretical and observational sentences. This evidential holism is not an accidental feature of our methods of hypothesis testing but, rather, one that follows with logical necessity from the logical structure of empirical confirmation and falsification. One does not have to accept the truth of evidential holism to see that it is an intelligible holistic position in favor of which there are plausible arguments.[45]

Given the familiarity of that type of holism, why has there been so much resistance to holism when applied to emergent phenomena? At least some of the opposition is a result of holism's perceived incompatibility with generative atomism and the accompanying suggestion that the remarkable clarity of the latter has to be replaced in some cases with a murkier set of representations. Indeed, a further fault of the slogan "The whole is greater than the sum of its parts" is that it suggests that, as emergentists, we must identify the source of the extra "something" that emergence gives us over and above what generative atomism provides—an unappealing task. Finally, in the case of the social sciences, political views about the importance of the individual have historically played a role in rejecting the existence of holistic properties.

To show that holism is not a necessary condition for emergence takes more work. Before undertaking that task, we must distinguish

45. One recalls that Quine's second dogma of empiricism was a form of epistemological reduction that he rejected while being an advocate of physicalism.

holistic properties from contextual properties.[46] First, I give a definition of "contextual" that is motivated by an opposition to immutability:

> *Definition* A property φ of an entity *a* is immutable with respect to contexts C and C* just in case φ is invariant across *a*'s presence in C and *a*'s presence in C*. An individual *a* is immutable with respect to contexts C and C* if and only if the essential properties of *a* remain invariant across C and C*.

Here, I take an essential property P of an individual *a* to be one such that if **a** lacked P, it would be a different type of individual. This version of sortal essentialism allows such things as electrons, helium atoms, and members of the type *Homo economicus* to have essential properties.

> *Definition* A property φ of an entity *a* is contextual with respect to contexts C and C* just in case φ varies across *a*'s presence in C and *a*'s presence in C*. An individual *a* is contextual with respect to context C and C* if and only if at least one essential property of *a* changes between C and C*.

Thus, *a* must have at least one essential property; individuals with none are neither immutable nor contextual. Contextualism allows a property to exist as a feature attached to a part of a system, the value of the property depending on the context, whereas nomological holism denies that the property can be attached to anything other than the whole system.

These definitions cover the special case where C is the empty context. Thus, if *a* has φ when in isolation (that is, when in C, the empty context) and φ is also possessed by *a* when *a* is present within some larger entity (i.e., C*), we have the situation often tacitly assumed in atomism. When at least one component individual of a system of individuals is contextual and the change in its essential properties is unpredictable, we have the kind of situation considered by Broad mentioned in section

46. Although contextualism is often associated with pragmatism, the view given here is ontological, not pragmatic.

3.1: "the characteristic properties of the whole R(A, B, C) cannot, even in theory, be deduced from the most complete knowledge of the properties of A, B, and C in isolation or in other wholes which are not of the form R(A, B, C)" (1925, 61).[47] Here, the context for the component A is the presence of B and C within the structure represented by R.

To keep the definitions simple, I shall violate one of the cardinal rules of philosophical wisdom and consider existence to be an essential property of an individual, so that if *a* exists within C but not within C*, then *a* is contextual with respect to C and C*. Note that in the case of properties, φ can have any adicity—monadic, binary, ternary, etc.—so that relational properties can be contextual or immutable.[48]

Contextualism need not involve downward causation. In cases where *a* is part of a system S, and C is the context consisting of *a*'s occurrence as part of S, contextualism can capture the influence of the rest of a system on one of its parts. But this need not produce either circular causation or reflexive causation, and so it is not subject to one of the perceived problems of downward causation. Peer pressure from some members of a group on another is a familiar example.

The concepts of a nomologically holistic property and of a contextual property are logically independent. Suppose that the nomological structure of a virtual economy is in part constituted by the rules that govern the society's currency. Then, the property of being able to purchase a beer in the Rusty Scupper is a nomologically holistic property of the totality of coins used for the purchase, which taken together sum to exactly the right amount—no part of that collection has the required property—but the value of the coins is not contextual with respect to that economic context and all other contemporaneous economic contexts that use the same currency; the coins have the same value whatever other transactions are made or whatever other coins and notes are around.[49] Note that there is nothing mysterious about this particular

47. For similar appeals to contextual features, see Stephan (1992, 1998).
48. Although some contexts can be represented by relations, in others the context does not have a definite adicity.
49. Of course, the coins only have purchasing power where they count as currency within an economy, but that is not the sort of contextualism at issue here. Also, the non-contextualism has a limited scope. If the quantity of money in circulation is drastically increased, the value of those coins will drop.

nomologically holistic property, which results from a straightforwardly additive process. This holism is a matter of nomological necessity but metaphysical contingency. A society with different economic rules could have replicated the effects of a progressive tax by making units of the currency subadditive; successive dollars added to the total have less and less positive purchasing power the more of them that are added, making their value contextual while leaving the total value nomologically holistic.

Conversely, contextualism can occur without nomological holism. Being a cause of measles is a contextual property of the measles virus, since it depends upon whether or not the context in which the virus is located consists of an individual who has been vaccinated, but the property of being a cause belongs to the virus, not to the entire biological system we call a human individual, hence is not holistic, and a fortiori is not nomologically holistic.[50] The reader can easily construct the other two cases required for logical independence.

3.5.1 Transformation Revisited

Having separated holism and contextualism, and continuing the theme of the previous section, I shall now argue in a more general way that one can take changes in the primitive elements as a way of understanding some cases of emergence, rather than by directly appealing to holism. In many cases, interactions between the primitive elements will be responsible for the transformations. We can capture this interactive aspect in the following way: Suppose you have a collection of fundamental entities $e_1, \ldots, e_n$ which, taken individually, can be considered to exist in isolation from one another. Those fundamental entities, as a result of a dynamic process, interact with one another and as a consequence of that interaction the entities diachronically change to $e_1^*, \ldots, e_m^*$ (where m and n may or may not be the same and at least one $e_i^* \neq e_j$ for $j \in \{1, \ldots, n\}$). The result is a system representable by $S(e_1^*, \ldots, e_m^*)$. The properties of the collective entity $S(e_1^*, \ldots, e_m^*)$ are synchronically determined

50. Casual talk may attribute the transmitting cause to an entire human, but the mode of transmission requires only presence of the virus.

neither by $e_1, \ldots, e_n$ nor by $S(e_1, \ldots, e_n)$, although we can suppose that they are determined by $e_1^*, \ldots, e_m^*$ and their relations. One might then be tempted to say that $S(e_1^*, \ldots, e_m^*)$ was something "over and above" the original elements $e_1, \ldots, e_n$ and their relations, and it would be this feature that motivated the belief that there was some holistic aspect of $S(e_1^*, \ldots, e_m^*)$. Yet as we have seen in the example of the mob, what one actually has is a diachronic transformation of the original components. In the type of situation considered, as a result of interacting with some or all of the other entities, at least one e_j is transformed into a different type of entity e_i^* and as a result has properties that are different from those e_j possesses in isolation. The properties of the system that has $e_1^*, \ldots, e_m^*$ as components differ from the properties that it would have, counterfactually and sometimes counterlegally, were it to be composed of $e_1, \ldots, e_n$. The interaction process is diachronic and cannot be captured by synchronic accounts alone.

Such interactions can help us to understand how some traditional approaches to emergence can be motivated. Take the simplest situation in which an individual e_i is transformed into a different type of individual e_i^* in virtue of interacting with the rest of the unchanged elements e_k, $k \neq i$, so that e_i is transformed but the e_k are not. Further, suppose that the transformation of e_i is not determined by the laws that govern the $e_1, \ldots, e_n$ either individually or in groups smaller than $e_1, \ldots, e_n$ but that the properties of the system $S(e_1, \ldots, e_i^*, \ldots, e_n)$ are determined by the $e_1, \ldots, e_i^*, \ldots, e_n$ and their interactions. When, as we saw in the quote from C. D. Broad above, the notation suggests that A, B, and C retain their identities within the whole, this places the burden of emergence on the relational aspects of the whole and fails to capture a key diachronic feature of the situation. This tacit appeal by Broad to the immutability of the components was and remains a result of the prevailing commitment to atomism. It also suggests that the key to emergence in such cases lies in either the transformation process leading to the production of the new entity e_i^* or in the interactions between e_i^* and the remaining entities.

This kind of novelty can be easily understood when we are considering transitions within a computational model, although it does not produce examples of ontological emergence. A red cell can transform into a green cell in a multistate cellular automaton without anything ontologically emergent appearing, even if the result of that transformation represents something novel in the world. The situation is less easy to understand with natural systems because it requires a new type of fundamental entity to be produced, having at least one novel property that behaves, or interacts with previously existing properties or other novel properties, in new ways. Epistemologically, an agent would have to collect data on repeated instances of such interactions in order to formulate a lawlike regularity, but ontologically, the way that the new property behaves and interacts is a fundamental and novel feature of the system, exhibited from the very first occurrence. The new lawlike regularity could apply to the system as a whole, and in that sense has some affinities with the view that emergence occurs when a certain level of complexity has been reached, but in other cases it could apply just to instances of e_i^*. An entity e_i does not have to interact with the system as a whole to be transformed into a different entity e_i^*. Iterated interactions with other individuals or subsystems, sometimes in conjunction with threshold effects, can achieve the transformation, or the transformation could be spontaneous. All of these considerations support the position that holism is not necessary for transformational emergence to occur.

Transformational emergence also suggests that we should keep separate two different cases of inferential emergence. In cases such as the appearance of traffic jams, the constituents of the system—the cars, vans, motorcycles, and other vehicles—are not transformed. What is unpredictable are the patterns of movement. In other cases, such as the appearance of a pattern in a multistate cellular automaton, the pattern exists in part because of transformations in the cell states at the fundamental level. Within the latter, the immediate reaction to the kind of situation just proposed is to note that if the dynamics of the process that leads from $e_1, \ldots, e_n$ to $e^*_1, \ldots, e^*_m$ are predictable and the system is structured by a quantitative relation S, the value of which is predictable from the intrinsic properties of the individuals $e^*_1, \ldots, e^*_m$ and the relations between them (arity $t \leq m$), then this is no case of inferential emergence. That reaction is correct. But in cases in which those dynamics are unpredictable because of the lack of a systematic theory of the processes involved, or because the complexity of the required calculations entails that we have to let the physical process work itself out to know the value of $S(e^*_1, \ldots, e^*_m)$, then we shall have a case of inferential emergence, often in the form of weak emergence in which we shall need to resort to step-by-step simulation of the dynamics.

There are many more complicated situations than the simple one I have just described, but the main philosophical points carry over. The appearance of a new law form is what writers tacitly seem to have in mind when writing about nomological emergence, but this is also unnecessary. An inverse square law holding between novel, emergent types of object and properties would be an emergent law just as much as one that involves a new kind of special function.

3.5.2 Contextualism and Compositionality

Because semantic contextualism has been so widely discussed in philosophy, some comparisons with this case are worthwhile, although I note that the semantic case is synchronic and is therefore an imperfect guide to the kind of contextualism that can occur in diachronic emergence. Standard examples of semantic contextualism are evaluative terms such as "heavy" for which the extension, and supposedly the meaning, changes from context to context. A backpack that is heavy to a toddler will not be heavy for her father. With a little work, many of these cases of contextualist meanings can be given a more stable content. "Heavy" means, roughly, "difficult or impossible to lift unaided," a definition that covers contexts involving me, ants, Olympic weightlifters, cranes, and many other uses. In such cases the meaning stays invariant, but the extension of the predicate "heavy" changes depending on the context. If it is troubling that this violates the commonly hold view that meaning determines extension, one can relativize the predicate so that it becomes "heavy for *x* at spatiotemporal location *l*."

Differently, some sentences can change their truth values when embedded in larger units: "This sentence is the last sentence of its paragraph" is always true in isolation, often false when not. Semantic contextualism is a feature exploited in literary tropes such as hyperbole, and there are many examples of contextual changes in meaning not only of individual terms but also of whole phrases. The phrase "hiking the Appalachian Trail" has, since 2009 in the United States, taken on the meaning of "visiting a woman who is not your wife," in addition to its usual meaning.[51] Which meaning is appropriate is generally determined by the larger context of its occurrence. Moreover, in its newer use, the meaning of the whole expression is no longer a function of the meanings of the constituents, whether they have their original meanings or not. In extreme cases, a single piece of syntax can have opposite meanings in two sentences, such as "He cleaved the log in two" and "He cleaved to his wife."[52]

Semantic contextualism is often referred to as "semantic holism," but this is only true in particular versions of the view. The difference between semantic contextualism and semantic holism is easy to see in the case of implicit definitions, in which a theoretical term such as "probability," takes on meaning only in virtue of the entire axiomatic theory within which it is embedded. For different theories of probability, such as those that do and do not endorse countable additivity, the meaning of the term "probability," as given by the respective implicit definitions, is different in the two theories, hence contextual; but the meaning remains a property of the term "probability" and not a property of the whole theory, even though it is determined by it. If this were not so, it would be difficult to separately attribute the meanings of the terms "probability" and "stochastic independence," both of which are given implicit definitions by the same theory. This makes the meaning of implicitly defined terms semantically contextual and not semantically holistic.[53] In contrast to implicit definitions, one of the virtues of providing an explicit definition for a term is that it provides a meaning for

51. The origin of the idiom was that in 2009, Mark Sanford, then governor of South Carolina, disappeared after announcing that he was going to "hike the Appalachian Trail." He resurfaced a few days later, and it was discovered that he had been in Argentina hiking the Appalachian Trail.

52. Of course, one could say that "cleaved" and "cleaved to" are simply different English terms, but these tactics are difficult to evaluate. In American English, "I wrote my mother" and "I wrote to my mother" are both considered to be syntactically sound and semantically equivalent sentences. Ordinary language is a shaky basis from which to draw philosophical conclusions. I hope the examples given here are free of those problems.

53. Although Beth's Definability Theorem shows that, for first-order logic, implicit definability entails explicit definability, this result does not hold within second-order logic, which is required to axiomatize such areas as probability theory. Butterfield (2011a, 2011b) has argued that Beth's Definability Theorem can be used to reconcile reduction and emergence within approaches to theoretical reduction.

the term in isolation. Using that definition, we can determine the meaning of a word independently of any context and that meaning remains invariant whatever its relations to words in other sentences.

Finally, in a case more directly relevant to our concerns, the rules for a cellular automaton are context-independent because they are the same whatever the state of the automaton, but the state to which a cell will transition is ordinarily dependent on the states of its neighbors. Two cells can be in the same state, yet because the states of their neighboring cells are different, the transition of those cells to their next state can be different.

Synchronic accounts of emergence rely heavily on compositional approaches to ontology. So, we must now distinguish contextualism from the failure of compositionality. Let φ be some property that is not nomologically holistic—that is, nomologically, φ can be possessed both by a system S and by S's constituents.

> *Definition* Let S be an entity having $S_1, \ldots, S_n$ as constituents. Then a property φ of S is compositional just in case the φ of S, the whole, is determined by the φs of the S_i.

I shall leave the exact sense of determination unspecified here since different determination relations, mathematical, functional, supervenience, causal, and others, can play this role. An elementary example of compositionality is when S is a country, the constituents are counties, and φ is the area. This compositional feature holds even when S is a scattered object, as when a country is composed of spatially disjoint regions such as an archipelago. The affinity of compositionality with generative atomism is obvious.

It is an often-implicit assumption of compositionality that the compositional property φ satisfies the immutability principle with respect to the constituents. That is, φ is a feature that is possessed by the constituents in isolation and φ remains invariant for each constituent within larger systems of which those elements are constituents. However, compositionality can hold in conjunction with contextualism, as the following example illustrates. Suppose that $a_1, \ldots, a_n$ possess properties $\varphi_1, \ldots, \varphi_n$ in isolation and that $\varphi_1, \ldots, \varphi_n$ are transformed into properties

$\varphi'_1, \ldots, \varphi'_n$ when $a_1, \ldots, a_n$ occur within the compositional context C′. Then, a property Φ of the whole system S can be given by $\Phi(a_1, \ldots, a_n) = f(\varphi'_1(a_1), \ldots, \varphi'_n(a_n))$, and compositionality is preserved when the determination relation is given by a function.

For a concrete example, suppose that the length of an object shrinks by one-half whenever it is spatially composed pairwise with another object, and that in isolation all objects have length 1. So, *a* and *b*, initially each of length 1, when composed have a combined length of $2 \times ½ = 1$; *a, b, c* when composed pairwise (in whatever order) have a combined length of $¼ + ½ = ¾$, and so on. The length of the composite is still determined by the length of the components, even though the lengths of the components are not immutable. This is not a case of emergence because no new property is involved.

It is more difficult to find examples of immutability without compositionality, but a simple example is $\sqrt{(2+2)}$, in which the values of the two token constituents 2 and 2 are immutable and the result is ∓ 2, the value of which is not determined by the values of the constituents of the functional expression above. Examples of the other two cases, compositionality with immutability, and the joint failure of both are obvious and so immutability and compositionality are logically independent properties.

3.5.3 Generative Atomism Again

Generative atomism, we recall, has at its heart the position that everything in a given domain results from composing properties possessed by immutable and individually distinguishable atoms, and that molecular entities are constructed from those atoms. It therefore requires that compositionality holds for the properties, and that immutability for the atoms holds within all contexts created by systems that are generated by the relevant rules or laws. So, following the suggestion that emergence occurs when generative atomism fails, let us consider situations when either compositionality or immutability fail for a system. In the case of individuals who become part of a mob, it is immutability that fails for the individuals. Similarly, Checkers World is a clear case in which immutability fails for the white and black particles. Both of these are cases of diachronic transformations. So, we have a reason to hold that an ineradicable failure of diachronic immutability, together with a change in at least one essential property, is a prima facie sufficient condition for emergence at the individual level. As we have seen in section 2.3, a failure of compositionality—or, in a more extreme way, the existence of fused entities—is also sufficient for emergence.

None of this is definitive of emergence as a general concept, but what we have in the kinds of examples we have examined so far is that the properties of the whole system are not determined by the properties of the components taken in isolation, even though they can be determined by the properties of the components when they are transformed into parts of the whole. In extreme cases, it is no longer possible to identify components of the whole, making the dependence of the latter on components moot.

One source of concern about the failure of the immutability principle might be that the method of analysis cannot be used to understand the system. Although the assumption is usually left implicit, analysis standardly assumes that the components of a complex system, when analyzed into entities separate from the whole, are the same entities as they were within the whole. This assumption underpins the methods of causal analysis and of experimental analysis. Experimental methods are designed to isolate individual causal influences on a system and unless a version of the immutability principle is satisfied by the causal factors, extrapolation from the experimental context to nonexperimental contexts will at best be difficult and the desideratum of external validity will be violated.[54]

Interactions can be managed if we know the transformation function that governs the change in the components between context C and context C′, because then the relevant inferences can be made. Given knowledge of the degree of compression of individual rubber blocks when they have similar blocks stacked on top of them, it is possible to derive the total height of the stack from knowledge of the thickness of

54. An experimental result has a high degree of external validity if that result holds in a large number of nonexperimental contexts and a low degree if it holds only in a small number.

the blocks in isolation. It is the lack of knowledge of the relevant transformation functions that constitutes a primary reason complex systems often exhibit epistemically and conceptually emergent features. One standard condition for a system to be complex is that there exist strong interactions between the components of the system, and this frequently entails that knowledge of the transformation functions cannot be obtained, so that the theoretical treatment must take place at the level of the entire system.

4

Inferential Emergence

4.1 A DEFINITION

We have already seen how the dynamics of processes play a role in various accounts of ontological emergence. In this chapter I shall discuss inferential approaches to emergence and the role played therein by diachronic features. The inferential approach takes as emergent the states of systems that fall outside the deductive or computational scope of some privileged theory. The theory may be privileged because it is taken to be fundamental, or to be unusually comprehensive. As C. D. Broad put it, "The only peculiarity of [a transordinal law] is that we must wait till we meet with an actual instance of an object of the higher order before we can discover such a law; and that we cannot possibly deduce it beforehand from any combination of laws which we have discovered by observing aggregates of a lower order" (Broad 1925, 79).[1] We see here the privileged position given to complete theoretical representations. If all phenomena within a given domain, including phenomena resulting from aggregates or other composite entities, can be predicted using a comprehensive theory of that domain, the theory will be inferentially complete and there will be no phenomena within the domain that are inferentially, and hence epistemically, novel. Conversely, if we have to resort to further observation or experimentation to know about the phenomenon at hand because that phenomenon falls outside the scope of what we previously considered to be a fundamental theory, then the phenomenon is novel with respect to that theory and so is a candidate for emergence. Broad's account was formulated before the development

1. See also Pepper (1926).

of completeness theorems and formal theories of computability, and it is, forgivably, unspecific about some aspects of what an ideal comprehensive theory might be.

More precise accounts of the relevant derivational relations underlie computationally based treatments of emergence, one of which—weak emergence—will be discussed in section 4.3.[2] Although inferential emergence can be, and often is, taken as a self-contained epistemological approach to emergence, I shall argue that it is less than satisfactory unless it accounts for how the lack of predictability results from the ontological structure of the system concerned. That this can be done by more recent versions of weak emergence is one of the significant advantages of that approach.

One way to capture the inferential approach is:

> *Definition* A state, law, or entity Z is inferentially emergent with respect to a theory, a model, or a simulation T having domain S if and only if (a) the status of Z makes it a prima facie candidate for inclusion in S, (b) T has the means for representing Z, (c) T includes rules or law statements that are considered fundamental to the domain S, and (d) it is impossible on the basis of T, together with specific conditions C representing fundamental facts about the domain S, to effectively derive or compute the representation of Z, where C entails neither the presence nor the absence of Z.

Note that this definition refers to nonrepresentational items such as the states and the domain of the theory, so that we already have some contact with ontology, but the emphasis is on the theoretical apparatus T. The terms T, Z, S, C, "impossible," and "effectively" serve as parameters in this approach that, when specified, produce a particular inferential account. The requirement that T includes law statements that are considered fundamental to the domain is imposed because a common characterization of an emergent property is that it is a nonbasic property that can be discovered only by empirical investigation, and it is

2. The work of Shalizi and Crutchfield (2001) on identifying efficient predictors for computational systems also fits into this tradition.

not epistemologically available solely by deduction from some set of law statements and facts that are considered to be fundamental to the domain in question. The fact that subsequent instances of the property could be predicted using inductive inferences after its initial discovery is compatible with the entity's being emergent with respect to T, because the inductive generalization used will not be among the original fundamental laws of T. We allow models and simulations in addition to theories because it is widely held that these two categories can produce inferentially emergent features, although in those cases, the links with ontological emergence are harder to make.

Derivability, and the lack thereof, has a wider scope than inferences about future states of systems. Because it is frequently believed that emergence can result when it is impossible to synchronically reduce one theory to another, we should keep open that possibility, at least within the tradition of Nagelian reduction.[3] Inferential emergence thus has synchronic and diachronic variants.

To avoid an inferential account from being too easily satisfied, it should be accompanied by arguments that S and the representational apparatus T have some claim to inferential priority (for example, that S is the physical domain) or to comprehensive scope and closure (that S is the domain of a well-established science such as molecular biology). A common approach is to restrict T to a representation of the parts of a system, their intrinsic properties, and the interactions between the parts.

The impossibility of effective prediction can be construed in two ways. The first is to take the impossibility as being impossibility in principle, using only the resources of T, as when there are undecidable sentences of T that represent states of the system. This is the approach taken by Gu et al. (2009). The second is to take the impossibility of effective prediction as referring to the unavailability of a computationally efficient derivation, such as the computational incompressibility of simulations appealed to in weak emergence (e.g., Bedau 2002), or a

3. E. Nagel (1961, 367) took one characterization of emergence to be the unpredictability of higher level properties from lower level properties. Nagel reduction will be addressed in more detail in chapter 5.

high degree of computational complexity as in Gu and Alvarado (2011). In the first option, the generative part of the generative atomism tradition breaks down because certain states of the system are underivable, given the synchronic and diachronic generating rules of the theoretical representation even when the ontology is one of wholes and parts. In other types of cases, such as complex adaptive systems, where the constituents of the system themselves are changed as a result of interactions between them, it is often the impossibility of predicting in practice over an extended series of interactions how the basic elements will change that is the reason for attributing emergent states to the system. In both of these cases, the emergence in question is primarily a kind of epistemological emergence, wherein some item of knowledge transcends the representational and predictive reach of an otherwise powerful theory.

A tighter connection with ontological emergence can be gained in cases where we have good reason to believe that T is a correct representation of S. If, for example, during the dynamical development of some or all of S, the original laws of S are inadequate to predict all future aspects of S—its states, its properties, its objects, or its laws—the need to resort to new empirical evidence to predict those aspects could just reflect an inability of our mathematical representations to fully capture the content of S. But if new, fundamental ontological features such as laws emerge, the predictive inadequacies would reflect the appearance of ontological diachronically emergent features.

If T is a final theory, of the kind discussed in connection with limit science, this may count as a reason to speak of the underivability as absolute, but logically, the attribution of emergence is still relational, holding between Z and T. This relational aspect means that, even in the non-limit case it can be an objective feature of the relationship between Z and T that Z is inferentially emergent from T, and even though later theoretical developments may produce a more sophisticated theory T′ from which Z can be derived.

The condition that Z is a prima facie candidate for inclusion in S is to prevent the emergence relation holding between independent domains—we do not want the properties of Mandarin Chinese to be emergent from the domain of electric eels. The phenomena of microphysics are not predictable on the basis of specifically biological theories, but that does not make spin emergent from biological properties because we do not expect microphysical phenomena to be derivable from biology. Perhaps this appeal to what is a prima facie candidate could be eliminated by reverting to levels, but we saw in section 3.3 that the levels approach has serious drawbacks. Moreover, we are dealing here with epistemological accounts of emergence, and the state of knowledge will usually fix both what is a plausible candidate for reduction and what is considered to be fundamental. Finally, deduction can fail because the ideal theory of the domain S does not contain the deductive resources to allow the relevant prediction to be made. One reason can be that there is no complete theory for D, where a theory T is complete for a domain if every truth about the domain can be deduced from T.[4] Completeness and incompleteness considerations are here taken in the sense relevant to logic and mathematics, and must be sharply distinguished from what was called "completeness" in the disagreement between Bohr and Einstein about the completeness of quantum mechanics (Bohr 1935; Einstein, Podolsky, and Rosen 1935). There, incompleteness was identified with indeterminism.

The case of indeterministic systems is covered by the definition because, in such cases, the state of the system fixes only the probability distribution of the outcomes, not the individual outcomes. Thus, the occurrence of an indeterministic event with a probability of less than 0.5, such as the decay of a radium-223 atom within the next second, is not inferentially emergent because it is possible to derive the probability of decay and nondecay from models of radioactive decay.

The definition relativizes inferential emergence to a theory or a model T, but requires T to contain rules or laws that are considered to be fundamental. This is motivated by the fact that a definition that allowed something to be emergent simply because we had not yet been clever enough to invent a theory to predict it is too weak.[5] It would allow the motion

4. For example, the strengthened finite Ramsey theorem is unprovable within Peano arithmetic. This theorem can be interpreted as being a part of combinatorial decision theory, and this subject matter would lead one to expect that the relevant sentence should be derivable, but there is no sense that our first criterion for emergence—that the emergent phenomenon has developed from something else—has been satisfied.

5. On the basis that axiomatic theories form some kind of ideal, we might want to make a distinction between systems such as classical mechanics, for which a set of axioms exists

of comets to be emergent before the introduction of Newton's celestial mechanics, but not afterwards. This kind of theory-relativity was the basis of the attitude toward emergence taken by Hempel and Oppenheim (1948), in which they claim that what counts as emergent is relative to our current knowledge base. This epistemic deficiency approach also underlies positions that characterize emergence in psychological terms and are rooted in the fact that emergent phenomena are often surprising (see, e.g., Edmund et al. 1999). The surprise approach is too subjective to be of more than pragmatic interest, but it can be accommodated in some cases. Recalling our general strategy that emergence occurs when an otherwise successful generative method fails, it is this failure that is often the source of the surprise, especially when the generative apparatus is considered to be fundamental. In other cases, it is the epistemic novelty produced by the appearance of a theoretically unpredictable state or pattern that lies behind the psychological surprise.

Hempel and Oppenheim's view toward emergence falls into a recognizable skeptical tradition in epistemology, and we must accept that there is always the possibility of discovering new types of representation that will undermine the unpredictability claim. Claims of inferential emergence that do not identify the theoretical apparatus with respect to which the emergence claim is made are thus always defeasible.

Finally, it is important to keep separate two kinds of claims: the first, formal claim that Z is emergent with respect to some representational device T, and the second, realist claim that T is a true and complete representation of the relevant domain. We can have formal grounds for asserting the first claim, while accepting that claims of the second kind are always fallible. Both constitute inferential approaches to emergence, but where there is substantial evidence for the second claim, we can link ontological and inferential approaches in a useful way.

In terms of our four criteria for emergence, an entity's novelty is construed as its unpredictability and the criterion that requires the emergent feature to result from something else is satisfied in terms of relations between Z and T. There is nothing in the inferential approach

but some predictions are impossible in practice, and systems for which not even fundamental laws exist—but I shall not pursue that aspect here.

that requires holism. The emergent state is autonomous in the sense that it lies outside the scope of T.

4.2 PATTERN EMERGENCE

Inferential emergence comes in synchronic and diachronic versions. As a general phenomenon, diachronic emergence can be considered either from the point of view of the models that generate the emergent phenomena, these models usually being computational, or from the perspective of the real systems. In this section I shall restrict myself to consideration of the computational models because the structure of those models is ordinarily better understood than are the real-world generating processes. Indeed, as noted in section 2.2, one advantage of models is that there is no room for speculation about the details of the examples. The algorithms that underlie them are explicitly given, thus avoiding the speculative mist that surrounds many discussions of emergence in the philosophy of mind. This advantage is also a drawback. The fact that the model is computational, where here "computation" is taken in the usual digital sense, entails that ultimately the output is the result of a process driven by generative atomism, even in cases where nonlinearity plays a central role. This makes deriving ontological conclusions from features of the computational model unconvincing. So, in looking to computational models for examples of inferential emergence, I place special emphasis on the emergence of patterns that are generated by these models, rather than just considering the processes that generate those patterns.

The difference between natural and computational systems is also important because questions about causation within computational models usually are inappropriate unless one adopts an extreme Humean approach to causation, one in which it is at best difficult, and in my view impossible, to distinguish between events that are causally linked and those that are merely associated. When it is necessary to discuss real systems, in most cases the context will make it clear if we are discussing the model or the material system.

Many standard examples of systems that are taken to exhibit emergent patterns, such as the Belousov-Zhabotinsky reaction (see section

1.8), develop those novel patterns as a result of the temporal evolution of the system. Nest building, hive thermoregulation, forming a spore-bearing structure, and mobbing predators are all examples of novel biological functions that result from end-state patterns, with benefits for the relevant biological groups including dilution of predation, group vigilance, and the inability of predators to visually lock onto one target in the group.[6] It is widely agreed within the complexity theory literature that these and similar patterns and structures count as examples of emergent phenomena. Although these claims should not be taken uncritically, they can serve as a plausible starting place for philosophical analysis.

I shall argue here for two theses about diachronic inferential emergence. The first is that the historical development of a system's dynamics is crucial to a given pattern's status as emergent. That is, it is not possible to determine whether a given token of a pattern is emergent simply by examining the pattern itself or its contemporaneous micro states. There can be two instances of exactly the same pattern, one of which is emergent and the other not, the difference being solely in the way in which they were generated. This shows that synchronic accounts of emergence are incomplete in the sense that the temporal development of a given state is often central to its status as emergent or not. This historical element is not a trivial consequence of the fact that temporal processes are involved, but evidence for the conceptually separate nature of diachronic and synchronic emergence. Whereas wet rot in wood involves a temporal processes but its existence can be determined on the basis of features present at a single point in time, it is not always possible to decide whether a property is diachronically emergent simply by examining the end product of the process. The second thesis to be argued here is that diachronic inferential emergence is about tokens of higher level patterns rather than types.

The structure involved in the emergent patterns need not be spatial structure because there are agent-based models in which only abstract spaces are included in the model and the social structure that emerges is nonspatial. One example is the emergence of social and intellectual

6. See Parrish and Edelstein-Keshet (1999).

networks in which, owing to the ease of long-distance communication, the geographical location of the members is largely irrelevant. Because of this, the previously cited example of bird flocking and many of the cellular automata models that are often used as examples of emergence are not necessarily representative of pattern emergence. Once we include these more abstract patterns, the development of structures as a result of the dynamics of a system is a common feature of computational models in many areas of science.

Let us call the development over time of a structured pattern in a computational or real system as a result of the dynamics of the system "pattern emergence."

> *Criterion* A sufficient condition for the presence of a pattern-emergent state in a system is the generation of a stable, nonrandom structure or pattern as a result of a self-organizing process.[7]

4.2.1 Nonrandomness

To take the conditions one by one, the demand that the pattern be nonrandom is imposed for two reasons. Consider a pattern that transitions via some process from one random configuration to another random configuration. This process violates the novelty criterion for emergence. Although random patterns are token distinct, qua random, they are type identical and so no new property has been produced in the transition. Imposing the condition also helps avoid trivializing the definition of emergence because the generation of a random pattern is too easy to achieve. For example, two-dimensional patterns can be generated on finite grids by running through the coordinates and randomly assigning a color to each cell of the grid; with very high probability, some pattern that is random will be present as the process continues. But if such things count as emergent, then so would the result of almost any random process, and that is too generous, violating the Rarity Heuristic.

7. This is a sufficient condition only, one reason being that stable structures considered to be emergent could result from processes that are not in the realm of self-organization.

Consider now a system that makes the transition from being ordered to being disordered. Should this be considered to display emergence, since it satisfies the qualitative novelty criterion? One reason to reject transitions from order to disorder as cases of pattern emergence, but to accept transitions in the opposite direction, is the apparent ability of the latter case to run counter to the general trend produced by the Second Law of Thermodynamics. This produces a prima facie novelty in the transition from disorder to order. Order emerging from disorder is unusual, hence novel in one sense, and it produces a prima facie demand for an explanation; disorder arising from order is commonplace, and so would not be considered as novel by many, despite the diachronic change from order to disorder. Viewed ontologically, this argument is unsatisfactory, and I do not present it as wholly convincing, even in the inferential realm, so perhaps the differential treatment reflects a cognitive bias.

A third type of possibility, where an initially structured state evolves into a different structured state, needs to be dealt with on a case-by-case basis because many systems will produce nothing novel, whereas others will generate genuinely novel features. This leaves the fourth possibility, the development of an ordered structure or pattern from an initially disordered state, as our main object of interest.[8]

4.2.2 Self-Organization

The basic idea behind self-organization is that structure develops in a dynamic system without interventions from outside the system except for the flow of some quantity, with this structure being the result solely of local rules operating on the constituents of the system. This structure

8. The requirement that the pattern be nonrandom is sometimes motivated by the belief that emergent phenomena occur in the region intermediate between completely random behavior and completely structured behavior, a region including the much publicized "edge of chaos." I shall ignore that motivation here. What it is to be nonrandom can be captured by some objective measure of computational complexity, such as Kolmogorov complexity. This measure should properly be called Kolmogorov-Chaitin-Solomonoff complexity, as it was independently discovered by Andrei Kolmogorov, Gregory Chaitin, and Ray Solomonoff, with crucial technical contributions from the first two. For brevity, I shall refer to the measure as Kolmogorov complexity.

is not accidental in the sense that were the system to be restored to its initial state and the microdynamics rerun, similar large-scale structure would, with a high degree of probability, reappear.[9] There is usually a large number of initial conditions from which the same final structure can emerge.

One feature that self-organizing systems need not have is the ability to change their structures in response to changes in external conditions. In contrast, one could think of *self-adjusting systems* as those (structured) systems that, without the need for external intervention, but in response to changes in external circumstances, can reorganize themselves to improve their fit to those external circumstances. The properties of self-organization and self-adjustment are independent, but one way to understand the connection between the two is to note that if the stable state is disturbed, the original mechanisms that gave rise to the self-organization are still in place and will restore the structure unless the system has been pushed out of the basin of one attractor into that of another. So, for example, Schelling's segregation model (see section 1.8) will result in resegregation if people are forced into a desegregated configuration, but are then free to move again.

The invariance of the structure under external perturbations can be produced by boundary conditions that shield the system from disturbances, as in a society with border controls preventing immigration and emigration. More interestingly, the system can retain its structure in the absence of such shields. Thus, we do not count the constantly shifting winds on a blustery day as autonomous objects, but we do count a tornado as a self-adjusting system because it retains its structure across time under the influence of crosswinds and interactions with houses. The structure produced by self-organization will sometimes be a property of a proper subsystem of the entire system, because the rest of the system may be needed for the self organization to manifest itself. For

9. Self-organization and related features are usually statistical properties of a system, and in many models, unusual initial conditions will prevent the usual structures from emerging. The characterization of self-organization is consistent with Haken (1988, 11): "We shall say that a system is self-organizing if it acquires a spatial, temporal, or functional structure without specific interference from outside."

example, the convection rolls in Bénard instabilities (see section 1.8) make sense only against the backdrop of stationary portions of the fluid, and with this, as with other dissipative systems, entropy must be transferred from the organized subsystem to the environment. The stable structures formed in one phase of self-organization can be used as the basic units in another phase of self-organization, thus producing a hierarchy of objects.

There is a different aspect of self-organizing systems—that of resilience under minor internal disturbances (damage) to the system. Here again, the temporal development of the system can be important. The system might be stable under small perturbations once its final structure is in place, but small changes in the environment early in its development might prevent the structure from developing further. For example, the fetal development process is sensitive to low levels of mercury, whereas adult humans can survive these levels. But it is again the persistence of the mechanisms that are distributed across the system that maintains the functioning of the entire system.

4.2.3 Generation and Stability

Turning to the generation condition of our criterion, there are two quite distinct roles that the dynamics of the microprocesses can play in pattern emergence. The first involves the dynamics that lead to the formation of the structured pattern. The second involves dynamics that provide the stability of the macro pattern as the microdynamics of the system continue. This stability leads to three different kinds of autonomy for the micro patterns.

In the first kind, the emergent macrostructure is stable under dynamic microprocesses involving the same constituents over time. We can call this *recirculating autonomy*. Two widely cited scientific examples of this type are Bénard convection cells and Couette flow (see section 1.8). The distinguishing feature of recirculating autonomy is the existence of a structure possessed by a fixed collection of entities having a stability period much longer than the period needed for the entities constituting the pattern to perform one cycle. A purely static pattern is a degenerate case of recirculating autonomy.

In the second kind of autonomous pattern, the macrostructure persists through changes in the micro constituents, a case that applies to standing waves in river flows as well as, in some cases, to the products of social insects such as ants and termites as members of the population die and are replaced by others. The standing wave persists even though countless different water molecules move through the three-dimensional region that contains the wave and the insects maintain the same architectural structure in the mound across many generations. The distinguishing feature of this second kind of autonomy, which we can call *replacement autonomy,* is the persistence of a stable structure as the original micro constituents are replaced by new individuals of the same type, this replacement usually occurring as a result of the dynamics at the micro level. Recirculating autonomy is a special case of replacement autonomy.

Some examples of replacement autonomy produce dynamical versions of the Ship of Theseus problem.[10] Universities have their populations of undergraduate and graduate students completely replaced every few years, and their faculty, staff, and administrators completely replaced over a longer period, but in some cases, especially when the institutional culture remains constant, we consider a university qua institution to have retained its identity across all these population changes. For some other kinds of entities, such as rock bands, the question as to whether they retain their identities when one member is replaced by another is more difficult to answer.

A third kind of system occurs when the macrostructure is stable under replacement of micro constituents from a specified class of equivalent types. We can call this type *equivalence class autonomy.* One example of this third kind of pattern emergence is the "invisible hand" of markets, where optimal pricing in markets results from many local interactions between economic agents; this optimization process is invariant across many types of agents provided that the constraint that the agents are rational is respected. Equity stocks are now more often traded by

10. The Ship of Theseus problem, a staple of undergraduate metaphysics classes, poses the question: If you replace the planks in a wooden ship one by one, is the end product the same ship as the one with which you began?

algorithms than by humans, but subject to certain constraints, statistical regularities in the equity markets remain stable.[11] A second example is patterns of activation in neural nets where artificial as well as natural nets possess the relevant functional structure. It is quite common, although not universal, for systems exhibiting recirculating autonomy to also possess equivalence class autonomy, and replacement autonomy is a special case of equivalence class autonomy.[12]

We thus have two distinct roles that the dynamics of microprocesses play in pattern formation and persistence. The first role involves the process that leads to the initial formation of the structured pattern. This is all that is involved in the generation of a stable pattern. The second role involves the persistence of the structure as the microdynamics of the system continue. This feature (normally preceded by the first) is present in all three kinds of structure invariance. The first of these roles is important for considerations of pure diachronic emergence, whereas the second introduces a different aspect of diachronic emergence—the persistence of an emergent pattern across time—that requires drawing upon concepts from both diachronic and synchronic emergence.

4.2.4 Pattern Emergence Need Not Be Discontinuous

With our criterion for pattern emergence motivated, we can see how it works for a specific example. I shall take a moderately complex case for purposes of illustration, but the assessment should scale up and down. The example is the Greenberg-Hastings model, which can be used to simulate basic Belousov-Zhabotinsky reactions, as well as neural excitation and relaxation modes. Consider a two-dimensional 256×256 14-state cellular automaton with von Neumann neighborhoods. A brief reminder—cellular automata are dynamical systems consisting of an N-dimensional array of cells capable of being in M different states. The

11. For an explanation and exceptions, see Humphreys (2014b).
12. There is a fourth kind of stability, discussed earlier in this section, that is sometimes considered to be important for self-organizing systems, one in which the pattern is stable under external perturbations. This is a feature that is characteristic of, for example, Reynold's boid models of the flocking behavior of birds. For details of the boid algorithms, see http://www.red3d.com/cwr/boids/.

state of any given cell is updated at each time-step by rules that depend only upon the state of that cell and the states of its immediate neighbors. The update rules for the Greenberg-Hastings model are:

1. If the target cell is in state *m* and an adjacent cell is in state 1, then the target cell goes into state 1. (Excitation of a cell by a neighboring excited cell)
2. If the target cell is in state *m* and an adjacent cell is in state $m + 1$, for $1 \leq m \leq 13$, then the target cell goes into state $m + 1$. (Relaxation of the cell)

Otherwise, the target cell stays in its current state.

From an initial random state displayed on the left in figure 4.1, the system evolves after 500 iterations into the pattern pictured on the right.

The pattern does not emerge until about 200 time-steps have been completed, and the full pattern takes about 500 steps to complete. Which initial random state occurs is irrelevant to the appearance of the final pattern. This example illustrates two things. First, some of the literature on emergence conveys the impression that as soon as a critical level of complexity is reached, an emergent phenomenon suddenly

Figure 4.1. Greenberg-Hastings model. 256 × 256 fourteen-state cellular automaton with von Neumann neighborhoods. Left: Initial random state. Right: Equilibrium state after 500 time-steps.

appears. Certainly, some emergent phenomena do result from discontinuities, such as first-order phase transitions, but the spiral-maze patterns only emerge gradually over time. It is an artifact of the emphasis on a few nonrepresentative examples that emergence seems to be an all-or-nothing property of a system, a view that has been fostered by discussions of consciousness, life, and other phenomena that have been taken to be dichotomous.[13] In being a graduated phenomenon, emergence is similar to self-awareness, which is plausibly a matter of degree, as is consciousness.

To assess this, we need to distinguish between the questions of whether an emergent state can arise from a continuous process and that of whether the final state can have degrees of emergence. The Belousov-Zhabotinsky reaction shows that the answer to the first question is yes. To address the second question, rather than using degrees of emergence, we can ask whether some states of a system can be partly emergent and others be fully emergent. Consider the set of 500 states between the initial and final states, inclusive, of the Greenberg-Hastings process above. One might deny that all the patterns exhibited in the 500 steps correspond to partly emergent properties by insisting that there must be a predicate in a specifiable meta-language that describes a given pattern before it can be a candidate for emergence. The initial and final states have simple perceptual and linguistic conceptualizations—random and spiral, respectively; the intermediate stages do not, so that an attempt to explicitly derive all the stages of the process at the macro level of conceptualization would not have access to the conceptual resources necessary for the derivation. However, this argument is subject to the charge of arbitrariness or of anthropocentrism, because a predicate can be introduced into a language as a primitive to refer to any arbitrarily selected pattern, and those that happen to already occur in languages tend to be those we humans find important. So, this attempt to deny gradualism about emergence is unsuccessful.

A second argument resists the idea that transient patterns count as even partly emergent by requiring that the stability conditions

13. Wimsatt (2007) contains a sustained argument for emergence not being a dichotomous property; see section 6.5 for more details.

contained in our criteria for autonomy be satisfied for a fully or partly emergent state to be present, thus preventing most of the intermediate states of the process from count as partly emergent. This seems convincing, if not overwhelmingly so. If we accept that argument, what we have is that the process leading to pattern emergence is gradual; but in this example, there are no states of partial emergence en route to the (fully) emergent final, stable state. That leaves open the possibility that in other examples, partly emergent states occur.

A final point is worth making about this example. Although it is often claimed that emergence is a result of increasing complexity in a system, the reverse happens in this example. As can be seen from the outputs of the model, the initial random state of the cellular automaton is more complex than the final, patterned state. This claim does depend on the definition of complexity used, but for the measure of complexity we have adopted, the Kolmogorov complexity is low for the generated pattern and high for the random initial state. This feature is consistent with emergentist traditions in which it is complexity that generates an emergent feature but the emergent feature can itself be simple.

4.2.5 Pattern Emergence Is Historical

We can now see why the historical development of a pattern is essential to its status as an emergent entity. It is an essential feature of emergence that the emergent entity emerge from something else. Consider a token of the spiral pattern that was generated by running the cellular automaton over N time-steps. The process by means of which that token is generated involves the iteration of the rules N times, and it is from that process that the pattern emerges. Now, suppose I print an exact duplicate of that pattern using a spiral-patterned rubber stamp. It is another token of the same pattern, but that token is not pattern emergent because it is generated instantaneously. In fact, that rubber stamp token is not emergent in any sense, because there is nothing from which it emerges; there is just the pattern itself, which was duplicated at a time instant. If the response is that the pattern emerges from the causal interaction between the rubber stamp and the paper, this will lead to allowing the products of most causal interactions as emergent entities, which many would find unappealing. It would also violate the Rarity Heuristic.

This reveals three things. Pattern emergence is a historical phenomenon—whether an instance of a pattern is emergent or not depends essentially upon the process that generated it. It is impossible to determine whether an array is pattern emergent by looking only at synchronic relations between the pattern and the spatial array of elements that constitute the pattern. Compare this with what is claimed about synchronic relations such as those of supervenience or realization. For those relations, one is supposed to be able to determine by examining an instantaneous state of a system whether the higher level property is emergent from the lower level. The reason for this lies in the fact that supervenience relations are about types, properties, and universals, as they must be for them to capture more than the token–token identity relations that both reductionists and anti-reductionists agree hold between the levels related by these supervenience relations. Furthermore, it is in the very nature of the necessitation relation embedded in supervenience relations that identical bases give rise to identical supervenient features. As a result, no synchronic account of emergence based on supervenience relations of which I am aware is capable of differentiating the two instances of the pattern.[14] This shows that a diachronic element to pattern emergence is necessary.

In addition, the argument shows that pattern emergence is about tokens or instances of patterns, not about types. In many discussions of emergence, the position is that if a given property is emergent in any of its instances, it is emergent in all of its instances. This, as we have just seen, is false. Exactly the same pattern type may be emergent in one instance and not emergent in another.

4.2.6 Properties of Pattern Emergence

Does pattern emergence implicitly appeal to levels through its reference to such things as macrostructure and microprocesses? No, because what

14. A definition of diachronic supervenience is given in section 6.2, but that does not help in avoiding the conclusion of our argument.

"macrostructure" means here is a global pattern, as opposed to local entities such as individual cells and their states. Patterns are essentially holistic features of a system—the spiral-maze property is a distributed feature of the entire grid or of a significant part of it—but as we saw in section 3.5, holism is not a sufficient condition for changing levels. There is thus no need for talk of levels here, except in a metaphorical sense that leads to no ontological distinctions.

Because any pattern that can be explicitly defined in terms of local properties has been reduced, diachronic pattern emergence is compatible with synchronic reduction. Because examples of ontological emergence are more controversial, it is less easy to find cases in which diachronic pattern emergence is accompanied by synchronic ontological emergence. Furthermore, this second possibility can only arise in noncomputational systems within which real, rather than simulated, interactions give rise to ontological emergence, and in these cases the pattern is not often a product of the ontological emergence. I offer, without great conviction, the example of ferromagnetism within which the alignment of magnetic moments counts as a diachronically emerging pattern that is the result of local exchange-energy interactions.

4.2.7 Multiple Instantiability and Multiple Realizability

Multiple realizability is often taken to be a feature that accompanies emergent phenomena, providing the basis for the autonomy of those phenomena. Here, I shall argue that the microdynamic stability of structure involved in recirculating autonomy or replacement autonomy involves multiple instantiability, rather than multiple realizability. We can discuss this in the context of patterns in cellular automata, but the conclusions generalize. Recall that replacement autonomy occurs when a specific type of cell is involved in the realization of the emergent pattern, but which individuals are involved is irrelevant as long as they are of the appropriate type. Consider the kind of stable pattern, illustrated in figure 4.2, that emerges in the Game of Life with binary cell states.

In this example, we have replacement autonomy as the glider moves across the grid and the pattern is instantiated at different time-steps by

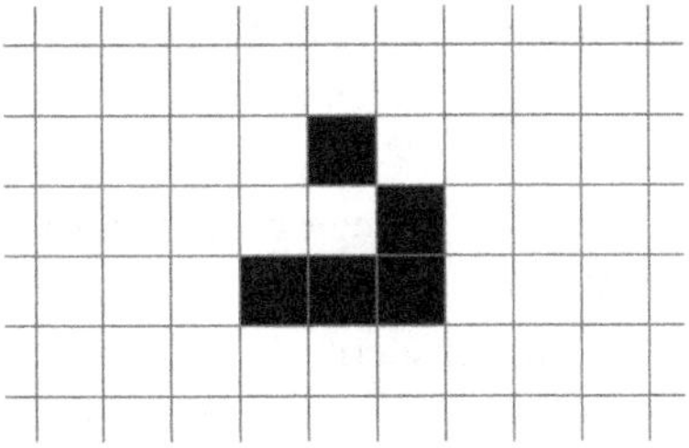

Figure 4.2. Game of Life glider.

different pairs of cells. Even though different cells are involved, they are always of the same type, and so multiple realizability is not involved. This shows that type–type identities are possible for replacement autonomy. A slightly different argument involving a pattern that oscillates between two sets of cells establishes the same conclusion for recirculating autonomy. This argument relies on a crucial assumption about pattern identity which becomes clear when we consider equivalence class autonomy. Suppose that instead of black and white cells, we have a 14-state cellular automaton, represented by fourteen different colors. Then, we need to have answers to questions such as: "Is the glider pattern present regardless of the individual cell colors that occur in a glider shape?" If the pattern is not color specific, then for this case of equivalence class autonomy not only do we have multiple realizability but also the position can be trivialized.

Instead of a glider shape, consider a quadruple of cells in a bow-tie pattern such as occurs at the top of the glider pattern. Any random array of cells will instantiate vast numbers of bow-tie patterns, as is easily seen from the initial random (left-hand) state of the Greenberg-Hastings algorithm. Unless we have an argument which establishes that a green-red bow-tie shape constitutes a different pattern from a blue-yellow bow-tie shape, in a situation in which two cells oscillate between a red-green configuration and a blue-yellow configuration, it will not be possible to have an objective criterion for whether we have recirculating or equivalence class autonomy. The bow-tie example is far too simple to represent most cases of pattern emergence, and the situation regarding pattern identity is, of course, much worse in other examples, such as the

Greenberg-Hastings case, not the least because we have no portmanteau term—no neat "bow-tie" terminology—to capture the pattern in these cases.

We are now in the midst of a tangle of familiar issues, none of which has an easy solution. We need to provide criteria for when a pattern is real, whether there are objective criteria that can be given for pattern identity, when a pattern counts as stable, and so on. There exist many attempts to address this problem. As a few illustrations, in Dennett (1991), the issue of what makes a pattern real is addressed through the design stance, the field of pattern recognition is devoted to such matters, complexity theory has grappled with the problem, and from its inception the field of statistics has developed objective criteria to replace the perceptual judgments we intuitively make about what counts as a regularity.[15] If I am correct about the role this kind of pattern persistence plays in diachronic emergence, then whether the criteria for pattern identity turn out to be objective, subjective, pragmatic, or based on some other ground, that feature will automatically carry over to computational diachronic emergence itself.

Thus far, I have not said explicitly in what sense these examples of patterns are emergent phenomena. We have the usual four features: novelty of the final pattern, its developing from an initial but different state by a process of self-organization, the fact that the pattern is a holistic feature of the entities that constitute it, and the autonomy of the pattern in any of the three senses described earlier. A second reason, beyond the satisfaction of our four criteria, is that many cases of pattern emergence fit the definition of weak emergence, which will be discussed in the next section, but there is an important difference between weak emergence and pattern emergence. Although in the latter, the diachronic process that led to the development of the pattern is important, the emphasis is on whether the end-state pattern has the appropriate kind of structure. In contrast, within weak emergence, any type of final pattern, random

15. An especially interesting recent attempt in information theory is Li et al. (2004). The kinds of patterns that occur in state space and that are treated in dynamical systems theory occur at a much higher level of abstraction than the geometrical patterns discussed here. For an interesting treatment of how such patterns are related to diachronic emergence, see Rueger (2000a, 2000b); also Strogatz (1994).

or structured, is allowed. What makes the end-state emergent in weak emergence is the nature of the process that generated it, not the features of the end-state itself.

4.3 WEAK EMERGENCE

Suppose that we take as a computational system any interpreted formal system that operates according to a deterministic algorithm in discrete time-steps. The processing can be either serial or parallel, but quantum computation and connectionist networks are excluded.[16] Then, anyone attempting to make sense of claims that computational systems produce emergent phenomena immediately faces a serious difficulty. The computations in such systems are, at root, combinatorial and everything is built from atomic primitives according to explicit rules, so they are straightforward examples of generative atomism. How can they then produce anything genuinely novel or emergent? We have discussed pattern emergence as one way in which this can happen. Approaches that fall under the general classification of weak emergence also have the potential to address this problem.

First, a terminology problem. It is unfortunate that the term "weak emergence" has often been used in the philosophical literature to denote any type of emergence that does not involve downward causation. That is, an entity is weakly emergent only if the causal effects of that entity can be identified with the causal effects of structured aggregates of the components of that entity, with those components existing at a lower level.[17] That characterization is too crude—the absence of downward causation by itself tells us nothing positive about what weak emergence

16. This definition also excludes some other computational systems, such as probabilistic automata, in order to avoid unnecessary complexities. For an excellent survey of theories of computability, see Shagrir (2016).
17. A representative use of this alternative meaning is: "Strong emergentists maintain that evolution in the cosmos produces new, ontologically distinct levels, which are characterized by their own distinct laws or regularities and causal forces. By contrast, weak emergentists insist that, as new patterns emerge, the fundamental causal processes remain those of physics" (Clayton 2004, 9).

is, and since it gives us only a necessary condition for weak emergence, systems that exhibit no emergent features even in the weak sense can also lack downward causation. I shall focus instead on a specific account of weak emergence due to Mark Bedau. I note here that Bedau himself explicitly denies the existence of downward causation (Bedau 2008, 450), although it is not a central feature of his account.[18]

Bedau's definition of weak emergence (1997, 2002, 2008) asserts that a state S of a system is weakly emergent from a state I of that system if S can be predicted from I only by an inferential process that reproduces all the intermediate states of the system between I and S. Our predictive access to this kind of system contrasts sharply with our access to future and past states of other types of systems, such as the solar system. A state consisting of a future solar eclipse is not weakly emergent from the current state of the solar system, because a prediction of the eclipse using classical mechanics can be derived without needing to know any of the intermediate states of the solar system.

Bedau's account has evolved over the years. In its original formulation, because of an appeal to underivability without simulation, it seemed to fall squarely into the inferential approach to emergence. In more recent formulations, Bedau (2008, 2012) has emphasized the idea of incompressible explanations and has provided arguments that certain natural systems exhibit weak emergence. In this version, weak emergence contains an element of ontological emergence that runs parallel to its inferential aspects. Because instances of weak emergence within computational models are easier to understand than examples in noncomputational systems, I shall start with the former, then move to systems that some of these computational systems represent. Weak emergence is often illustrated using the canonical system of a two-dimensional cellular automaton (CA), but it is better to think in terms of a broader class of simulations because the spatial relations that are present in the standard representations of cellular automata can be misleading.

The core feature of systems exhibiting weak emergence is that the process leading to the end-state of the system is computationally, and hence predictively, incompressible. That is, the only way to predict the

18. A predecessor to Bedau's approach can be found in Darley (1994).

behavior of such a system is to simulate it using a step-by-step iteration of the dynamical process that leads to the emergent state. Roughly, the idea is that many traditional theoretical predictions can "jump over" all or most of the intermediate states of a system to derive the state of interest, whereas a step-by-step simulation must go through all those intermediate states. The philosophical motivation for accepting this criterion as having captured a certain kind of emergence draws on the tradition that emphasizes the essential unpredictability of emergent phenomena on the basis of ideal theory. As a historical point, Laplace's famous account of determinism seems to deny the existence of weakly emergent properties, since Laplace's ideal intelligence has the feature that at any given instant: "the future, as the past, would be present before its eyes," rather than having to wait for calculations of various lengths to be carried out.[19] One can excuse Laplace for having thought that all systems would yield to the compressible computations that Newtonian celestial mechanics exemplified.

We can use traffic as a canonical example. Suppose you are in a car on a controlled-access road, on which light traffic is flowing freely. You know exactly where you are, where you are going, and your cruise control is maintaining a steady 60 mph. You can thus predict, now, using an elementary kinematical model, that, unless something extraordinary occurs, in exactly one hour you will be in Charlottesville, Virginia. Moreover, the number of computational steps required to make the prediction is largely independent of how far in the future the predicted event will occur.

Contrast that situation with one in which you are in a car trying to leave the pasture used as a parking lot at a rock music festival. There is only one exit, traffic moves in a stop-and-go fashion, you must merge into other lanes of traffic periodically, you must rely on other drivers to let you enter those lanes, and some of those drivers are in an altered state

19. More fully, "Given for one instant an intelligence which could comprehend all the forces by which nature is animated and the respective situation of the beings who compose it—an intelligence sufficiently vast to submit these data to analysis—it would embrace in the same formula the movements of the greatest bodies of the universe and those of the lightest atom; for it, nothing would be uncertain and the future, as the past, would be present before its eyes" (Laplace 1902, 4).

of consciousness. There are clearly defined time-steps in this stop-and-go situation, and at each time-step, some interaction with at least one other car takes place. Which state you will be in one hour from now—the state being your position and velocity—is (from your perspective inside the system) knowable only by going through all the intermediate states between now and one hour hence. And, a fortiori, this is true for the state of the system as a whole. (We assume the system is deterministic so that there is a fact of the matter about what the future state is.) If one were trying to predict the future state using a traffic model, predictions would usually be generated by using an agent-based simulation to successively calculate each stage in the iterated interactions. It is this need to go through this step-by-step representation of the states of the system that is characteristic of weak emergence, in both the system and any model of it.[20] Weak emergence is an illustration of the importance of diachronic emergence. Traditional mathematical models must be replaced by running computer simulations over time, on concrete computational devices, to replicate the temporal development of the system.

Bedau originally defined weak emergence in the following way: "Assume that P is a nominally emergent property possessed by some locally reducible system S. Then P is weakly emergent if and only if P is derivable from all of S's micro facts but only by simulation" (2002, 15). According to this definition, it is properties of the system that are weakly emergent rather than states, but for expository convenience I shall use the latter here. A state being an instance of a property, it is the derivability of the associated state that is the real focus of weak emergence.

Here is some elaboration: A nominally emergent property is one that can be possessed at the macro level, but is in principle incapable of being possessed at the micro level. What counts as the micro level and the macro level is largely conventional. A nominally emergent property is thus what Broad (1933, 268) called a "systemic property."[21] As

20. This example is not a case of the more commonly cited example of a traffic jam emerging from previously free-flowing traffic.

21. See Stephan (1999) for a discussion of this idea.

an example of nominal emergence, the only shape that individual cells in a cellular automaton can possess is that of squareness, and thus the property "is L-shaped," which can be possessed by clusters of cells but not by individual cells, is nominally emergent. Within the context of weak emergence we can take the relation between the two levels of properties as being grounded in a compositionality criterion. Nominal emergence does not always involve conceptual emergence because, as in the case of the "is L-shaped" predicate, the nominally emergent predicate can sometimes be defined in terms of lower level predicates and relations.

Although nominal emergence was used by Checkland (1981) as a criterion of emergence, and one often encounters casual remarks to the effect that emergent features are those that could not be possessed by systems at a lower level, it is too weak to capture what is important about emergent phenomena. Here is one reason why: the atoms of generative atomistic systems are noncomposite. This makes the property "being composite" nominally emergent relative to the fundamental level, and so if nominal emergence were sufficient for emergence, this would automatically attribute an emergent property to any aggregate collection simply because it was composite. That is too generous.

Turning next to local reducibility, a locally reducible system is a system in which all of the macro properties are structural properties. This means that the state of a micro entity consists of its location and intrinsic properties, whereas the state of a macro entity is simply the aggregate of the states of its micro constituents together with their (usually spatial) relations.[22] Weak emergence is, therefore, compatible with synchronic reduction in the sense that a locally reducible macrosystem is ontologically reducible to its constituents and their spatial relations. This puts weak emergence into the category of positions for which emergence is compatible with synchronic reduction, a category that includes Wimsatt's (2007) non-aggregativity approach.

A simulation in the special sense used by weak emergence—and it is a special sense—is a step-by-step process that replicates the time development of the system at the micro level. Adapting the approach

22. Other, less restrictive definitions of structural properties are given in section 6.6.

in Rasmussen and Barrett (1995), the computational incompressibility aspect of weak emergence can be represented in the following way:[23] Consider the kinematics of a system S that is modeled as a temporally discrete process. An updating function U takes the current state of the process and transforms it into a future state. There is, then, an essential difference between these two cases:

$$S(t+T)=U\left[S(t),t+T\right], \text{ for arbitrary } T, \qquad (4.1)$$

and

$$S(t+T)=U_T\left[U_{T-1}\left[\ldots\left[U_1\left[S(t),t+1\right]\right]\ldots\right]\right] \qquad (4.2)$$

where U_i is the updating operator at timestep i. In the usual case where $U_i = U$ is fixed, we have $S(t + T) = U^T[S(t), t + T]$.

The situation represented by equation (4.2) is the completely computationally incompressible case, whereas equation (4.1) is the completely computationally compressible situation. It should be clear that there are many degrees of compressibility between the cases described by (4.1) and (4.2), and weak emergence therefore comes in degrees. (Bedau 2008, 447; see also Assad and Packard 1992). This then raises the question of the appropriate measure of the degree of compressibility involved. One option is to use Kolmogorov complexity, but this captures the wrong aspect of computational incompressibility. To see this, distinguish between two logically independent measures: the length of the shortest algorithm that produces the given sequence of states, which is the feature that the Kolmogorov complexity measures, and the number of states intermediate between the initial and final states that need to be computed, which is the feature connected to weak emergence. It is possible to have a very short algorithm that can compute all those intermediate states. Conversely, one can have a very long algorithm that can get from the initial to the final state directly. Kolmogorov complexity is thus a poor measure of weak emergence, because the process leading

23. Another caution: some of the physics literature uses "undecidable" as a synonym for "computationally incompressible," the former a use that can lead to serious ambiguities for those familiar with recursion theory. So, it is best to stay with "computationally incompressible."

to a weakly emergent state can have a very low Kolmogorov complexity while being highly incompressible.[24] (For alternative measures, see Hovda 2008.)

4.3.1 Illustrations: Bird Flocking and Traffic Jams

The flocking of birds and the occurrence of traffic jams are frequently cited as examples of emergent phenomena. What is less common is an explanation of why they should count as emergent. Although there are many fascinating aspects of bird flocking and the associated phenomenon of fish schooling, I shall address here only the property of spatial cohesion over extended periods of time. The flock, which consists of the individual birds that are its members, has the remarkable ability to not undergo dispersion or collapse over time. In computational models of flocking, this spatial cohesion is a result of the individual motions of the birds satisfying three simple rules of separation, alignment, and attraction. These are, respectively, individuals maintaining a minimal separation from their nearest neighbors, steering in the average direction taken by those neighbors, and not exceeding a maximal separation from their neighbors. The satisfaction of these rules leads to the property of spatial cohesion resulting from the motions of the individual birds; it is holistic in the sense that it is a property of the group as a whole (and also by some subgroups) and cannot be possessed by individual members of the flock. Its cohesion is a novel feature of the aggregation of birds. Finally, the flock property is preserved under the substitution of individual birds from the same species, giving us replacement autonomy, so that all four of our general criteria are satisfied.

In what sense is this an example of weak emergence? At the level of the motions of the individual birds, predicting the exact state of each at time T and aggregating to arrive at the specific state of the flock at T can be done only by simulating within the model the iterated sequence of temporally discrete interactions between the birds. There is no shortcut

24. The exact value of the Kolmogorov complexity of a sequence is an uncomputable function, and this raises difficulties for proving whether a given process is incompressible.

at this level of representation to predict the states of the flock other than by running the step-by-step simulation. Of course, once the new concept of a flock has been introduced, and it has been established over the course of many simulation runs with different initial conditions that the formation and preservation of flocks is a robust feature of the model, a new theoretical representation can be used to make those predictions that allow a compressible prediction of stability.

In section 2.1, I argued that mobs probably do not exhibit ontological emergence. However, they, or models of them, do give rise to weak and conceptual emergence. Weak emergence occurs in models of mobs because of the inability to predict some future states of the mob, such as its level of violence, without resorting to a step-by-step simulation. One reason for this is the rapidly changing properties of the component agents, a feature that can be modeled within a complex adaptive systems approach, and the lack of a comprehensive theory about the transformation function that results in changes in the members of the mob. There is an essentially diachronic aspect as the changes in individuals spread through the system, and once that has occurred, compositionality is all that is required to arrive at the holistic property of group violence. It is interacting with like-minded people that allows the spread of anger from a few individuals to most of the members of the mob. Once that dynamic has happened, there is no mystery to the holism: it is because the individual members of the mob have changed and compositionality can aggregate the increased levels of individual anger to a high level for the whole group. There is no mysterious "downward causation" here; there are simply interactions at the agent level and compositionality. There will also be conceptual emergence, leading to the formulation of group-level law statements, in the sense that the term "mob" is employed in sociological theories as a sui generis predicate.

From the perspective of our taxonomy, the computational states fall squarely under the definition of inferential emergence given in section 4.1, if there is no predictively more efficient function available that will render the predictive process compressible. There is thus usually an epistemic element to the claim that P is a weakly emergent property, even if the definition itself is non-epistemic, because we are relying on a lack of knowledge of whether or not a more effective predictive technique exists. In some cases the nonexistence of such a method can be proved, but in many cases it cannot. We can see this with the following example.

Consider calculating the Nth digit of the transcendental number π. For centuries, any finite initial segment of π was assumed to be computationally incompressible, in the sense that the only way to calculate the Nth digit of π was to first compute all the preceding N-1 digits, making the sequence computationally incompressible. Then, in 1995, Simon Plouffe discovered an algorithm for directly calculating the Nth digit.[25] (Such algorithms are known as "spigot algorithms.") This immediately makes the sequence computationally compressible.

4.3.2 Assessment

Weak emergence succinctly captures a great deal of what is considered emergent within the fields of dynamical systems theory and complexity theory, but one aspect of it bears examination. As Cyrille Imbert (2006) has noted, Bedau's definition places no restriction on the kind of end-state property that can weakly emerge from a simulation. Imbert argues that this is a defect in the definition, although it should be noted that Bedau was aware of this consequence of his definition and addressed it in his 1997 article. As he wrote:

> One might object that weak emergence is *too* weak to be called "emergent," either because it applies so widely or arbitrarily that it does not demark an interesting class of phenomena, or because it applies to certain phenomena that are not emergent. For example, indefinitely many ad hoc Life macrostates are (for all we know) underivable without simulation. . . . But this breadth of instances, including those that are arbitrary or uninteresting to "emergence

25. The algorithm is now known as the Bailey-Borwein-Plouffe algorithm. For details of the algorithm, see Bailey, Borwein, and Plouffe (1997), and for a discussion of philosophical issues related to this example, see Dubucs (2006). Dubucs has emphasized that in the case of weak emergence, the key thing is not predicting "ahead of time," because with many weakly emergent systems, the computer simulation could do it faster than the real system; it is the number of steps involved.

theorists," is not a problem or flaw in weak emergence. (Bedau 1997, 394)[26]

What is at issue here is whether we should distinguish between novel future states that are weakly emergent and weakly emergent states that have no interestingly novel features. It is thus worth exploring in a little more detail whether we need to further refine the concept of weak emergence by also requiring that a weakly emergent state have some identifiable synchronic structure. I shall argue here that we need to maintain a sharp separation between our diachronic conceptions of emergence and our synchronic conceptions of emergence, because the diachronic motivation of weak emergence is very different from what drives most philosophical accounts of synchronic emergence.

Recall our criterion that for a property to count as emergent it must be novel. What counts as novel differs among different accounts of emergence, but one thing is fairly clear: whether a given property instance E is novel or not has to be decided with respect to property instances occurring within the process that leads to E, and not with reference to whether similar instances have occurred in any other process. To require that a property instance be absolutely new is too strict, for then only the first ever instance of a property would count as emergent. Having noted this, we can then give a second argument for our earlier conclusion that pattern emergence concerns tokens and not types (see section 4.2.5).

Consider a process within which weak emergence takes place over the first N time-steps, the (N + 1)st state being a token of a novel property type P, where P has some identifiable structure. The system's rules then maintain that state for an unbounded number of further time-steps in a way that is predictable in a highly compressed way. In such a system, the (N + 1)st state is weakly emergent but no further instances of P count as weakly emergent because their prediction involves compressible processes. Contrast this with a system having a different set of

26. Bedau has confirmed (personal communication) that he was aware of this aspect of weak emergence when he wrote his 1997 paper. In contrast, Rasmussen and Barrett (1995) require new structure at a higher level for emergence.

rules which evolves into a state of type P via a computationally incompressible process of length N, and in which another token of property type P is generated by a further computationally incompressible process after another M time-steps. No states of type P occur between the N + 1st and (N + M + 1)st time-steps. In this case, both instances of P are weakly emergent. It is the candidate state that is generated and its history that is important, and in these cases one has to examine the entire cross-temporal process in order to classify the state as emergent or not.

This reinforces our earlier conclusion that the criteria for identifying instances of diachronic emergence are partly diachronic. However, these considerations do not resolve the question of whether one should restrict the class of weakly emergent states in some way in an effort to combine the diachronic and synchronic approaches to emergence, and this remains as an open problem in diachronic weak emergence.

4.3.3 Weak Emergence as Explanatory Incompressibility

Although weak emergence was originally defined in terms of computational simulations, it is also said to occur in the modeled systems themselves.[27] The idea is that in natural systems within which components interact repeatedly through local causal relations, the computer simulation directly mimics the step-by-step process by means of which the natural system produces the final state that is considered to be the emergent feature. Examples cited by Bedau include states of traffic jams, states in protein folding processes, and states of certain kinds of artificial life systems.[28]

Here, I want to argue that although he is correct that weak emergence is not "just in the mind," there is a conceptual aspect of weak emergence that cannot be eliminated. In his 2008 formulation, rather

27. As Bedau puts it: "The phrase 'derivation by simulation' might seem to suggest that weak emergence applies only to what we normally think of as simulations, but this is a mistake. Weak emergence also applies directly to natural systems, whether or not anyone constructs a model or simulation of them" (2002, 17). See also Bedau (2008).
28. Because these systems involve continuous variables, the "not explainable except by simulations" must involve an appeal to approximations.

than relying on underivability except by simulation, Bedau characterizes weakly emergent phenomena as those having macro states that are explanatorily incompressible. "If P is a macro-property of some system S, then P is weakly emergent if and only if P is explainable from all of S's prior micro facts but only in an incompressible way" (2008, 442).

In order to understand what is being claimed, it is useful to draw a distinction between two kinds of computer simulations. In what we can call *direct simulation*, the structure and dynamics of the simulation models are similar to, perhaps even isomorphic to, the structure and dynamics of the modeled system. Agent-based models are frequently of this kind. In what we can call *indirect simulation*, this similarity is absent; the representation used in the simulation uses mathematical syntax that, perhaps because of the use of approximations, idealizations, and limits, does not directly reflect the structure and dynamics of the real system.

Bedau's conception of an explanatorily incompressible explanation appears to most easily be applied to direct simulations, and it is that conception which underlies the claims about weak emergence as a property of natural systems. The idea is that in natural systems within which components interact repeatedly through local causal relations, the computer simulation directly mimics, and has to mimic, the discrete, step-by-step process, by means of which the natural system produces the final macro state that is considered to be the emergent feature.

This representational aspect cannot be eliminated from weak emergence, even in its objective form.[29] Consider the two traffic examples described at the beginning of this section. In the actual world, your car has to travel through each intermediate time point in order to reach the end-state in both the highway scenario and the parking lot scenario, and this is true for any real system. It is only from a theoretical or modeling perspective that compression makes sense. You cannot, time dilation aside, eliminate steps in time or compress temporally extended concrete causal processes, as opposed to our representations of them. A major question is, then, how we can eliminate the possibility that there is an alternative representation of a weakly emergent process that allows a compressible computation. In the mathematical realm, the possibility

29. Bedau (2008, 15) agrees with this assessment.

of discovering an alternative method that allows intermediate stages between the initial input and the stage desired is illustrated by the discovery of spigot algorithms mentioned earlier. More generally, unless there exists a proof that the algorithm on which the weak emergence claim is based is optimal, we must resign ourselves to the fact that a claim of weak emergence is always defeasible and dependent upon the representation used.

4.3.4 Weak Emergence and Explanation

Gregory Chaitin, one of the inventors of algorithmic information theory, has written that understanding is compression.[30] If this is true, then we do not have a theoretical understanding of instances of weak emergence. Consider again the parking pasture scenario. In that situation, maximally incompressible explanations result from the fact that each time-step involves an interaction between cars, and every such interaction must be modeled in order to capture the predictively relevant causal interactions in the system. This need to comprehensively simulate the process can be understood in the light of Salmon's (1980) causal process model of explanation. In that approach, understanding is conveyed through knowledge of the causal interactions and propagations that led to the outcome, and it is knowing the structure of the world itself, rather than our theory of it, that provides understanding. In the kind of extreme situation such as the parking pasture, not only is there is an interaction at each time-step but each causal interaction can be known only by being acquainted with it, rather than its being predictable from other interactions of the same type. The only causal propagations are the single time-step movements of cars. Almost all such explanations will be so complex as to be unintelligible to humans, unless a reconceptualization of the dynamics at the level of macro properties can be provided. Compare this to the situation on the highway. There are few causal interactions between the initial and final states, and this simple causal structure allows us to straightforwardly gain the knowledge of the system we need to understand it without the need to reproduce the entire causal process.

30. Chaitin (2005, 6).

In some systems, moving up to the macro level does produce a significant increase in predictive and explanatory economy. Take the computer cellular automaton called the Game of Life. There, the appearance of macro-level patterns that are the result of complex micro-level interactions, such as the R-Pentamino pattern, cannot be effectively predicted except by running the simulation. But once their existence has been established, the patterns are stable enough that the rules describing their subsequent behavior can simply be described by an inductive procedure at the macro level using the new concepts describing the macropatterns.

Is there some other sense in which weakly emergent states are explainable? Consider Philip Kitcher's account of explanation. Although he is best known for promoting the unification account of scientific explanation, embedded in Kitcher's theory is an interesting distinction:

> Hempel's proposal that explanations are arguments appears to embody this conception of arguments as *premise-conclusion* pairs. But, on the systematization account, an argument is considered as a derivation, as a sequence of statements whose status (as a premise or as following from previous members in accordance with some specified rule) is clearly specified. An ideal explanation does not simply list the premises but shows how the premises yield the conclusion. (1989, 431)

This distinction is important. Because of the computational incompressibility of weakly emergent processes, the computational equivalent of the detailed sequence of explanatory deductions can be obtained only by directly reproducing the steps in the physical process. In many cases, understanding all these intermediate steps is impossible for humans. This is the feature that I have elsewhere (Humphreys 2004) called "epistemic opacity," and it explains why some physical states have been assessed as inferentially emergent, even though they are predictable not only in principle but also in practice.

In situations in which the inferential steps from the initial state to the final state are relatively simple, once the appropriate conceptual apparatus has been introduced, the details of the intermediate steps are unnecessary for understanding why the explanandum sentence is true. In the parking pasture example, you know that you will not exit in less than five minutes and probably not in more than twelve hours, but between those two extremes the time to exit is highly path dependent and unpredictable. In contrast, not only does one not need to know the exact thermodynamical path that leads to a phase change in ferromagnetism in order to understand why the phase change occurred—a generic account of how spin flips occur in a lattice as the temperature approaches the critical value will suffice—it would be explanatorily counterproductive to describe such a path. In a related way, once a proof by cases had been laid out as part of the proof of the Four Color Theorem, the combinatorial details that underlay the computerized checking of each case were irrelevant to human understanding about why the overall conclusion follows. All that is important is that it is true that the relevant lemma holds for each case and then algorithm reliability issues are what counts.

5

Conceptual Emergence

> [I]f you knew everything about water molecules and you had a computer good enough to follow how every molecule in a glass of water moved in space, all you would have would be a mountain of computer tape. How in that mountain of computer tape would you ever recognize the properties that interest you about water, properties like vorticity, turbulence, entropy and temperature? There is in the philosophical literature a term, emergence, that is used to describe how, as one goes to higher and higher levels of organization, new concepts emerge that are needed to understand the behavior at that level.
>
> Stephen Weinberg 1987, 434

5.1 CONCEPTUAL INNOVATION

Ontological and inferential emergence are often accompanied by conceptual innovations. This is especially common when the sheer complexity of a compositional system's states forces a redescription of those states using a new vocabulary. I shall argue here that conceptual innovation has played an important role in the development of ideas about emergence, that conceptual innovation alone does not itself constitute a distinctive form of emergence, and that, as with other topics we have examined, the role of conceptual innovation has been distorted by an emphasis on the synchronic cases.

We can begin with one of the most influential articles on emergence written in the twentieth century, one that, as with Mill's nineteenth-century work, never mentions the word "emergence." The great insight that Philip Anderson presented in his article "More is Different" (1972) was to focus on what happens when we try to actually carry out the construction of a higher level theoretical apparatus using only the theoretical and conceptual resources of some preferred lower level. Anderson had come to realize that even if ontological reduction were true, in the sense that all that the world contained was an enormously complex collection of interacting entities governed by fundamental laws—and Anderson insisted that "we must all start with reductionism, which I fully accept" (394)—the scientific project of constructing a usable theory of large-scale systems from the theories governing their constituents would often be practically impossible.[1] "The ability to reduce everything to simple fundamental laws does not imply the ability to start from those laws and reconstruct the universe. . . . The constructionist hypothesis breaks down when confronted with the twin difficulties of scale and complexity. . . . At each stage entirely new laws, concepts, and generalizations are necessary" (393). As a result, new concepts had to be introduced for the new domain in order to effectively predict and represent the phenomena in that domain. It was no coincidence that Anderson worked in condensed-matter physics, which studies the collective behavior of large ensembles. His interest in emergence was prompted by phenomena that violated symmetry principles which were universally respected at more fundamental levels.[2] In the intervening years, the correctness of Anderson's insights has been amply confirmed.

Despite the widespread influence of Anderson's article, explaining what its author meant is not straightforward because the article, for

1. Anderson's attitude toward reduction was anticipated by Paul Dirac. In his 1929 paper 'Quantum Mechanics of Many Electron Systems," Dirac wrote: "The underlying physical laws necessary for the mathematical theory of a large part of physics and the whole of chemistry are thus completely known, and the difficulty is only that the exact application of these laws leads to equations much too complicated to be soluble" (714). Dirac differed from Anderson in seeing the future as one of developing approximation methods, rather than as one of developing autonomous levels of theory. Thanks to Johannes Lenhard for drawing my attention to Dirac's paper.
2. During the debate about whether to construct the Superconducting Super Collider (SSC), Anderson and Steven Weinberg held conflicting views on whether the cost of the project was a wise investment of scientific resources. In terms of reduction, the views of Anderson and Weinberg, at least according to their explicit publications on the topic, are not very far apart.

understandable if regrettable reasons, often indiscriminately moves back and forth between material and linguistic modes of expression, between talk of properties and talk of concepts, and between talk of laws and talk of the theoretical descriptions of laws.[3] It is easiest to see what Anderson was advocating by considering his position on laws or, more accurately, law statements. For him, "The reductionist hypothesis . . . [is the thesis that] all of the animate and inanimate matter of which we have any detailed knowledge, are assumed to be controlled by the same set of fundamental laws, which except under certain extreme conditions, we feel we know pretty well" (Anderson 1972, 393). Although the reference to "controlled by . . . laws" suggests a realist attitude toward fundamental laws as, so to speak, directing the development of the system, the nonfundamental laws, if there are such, seem to have a different status, as the reference to "entirely new laws, concepts, and generalizations . . . requiring inspiration and creativity" above indicates. This reference to concepts, generalizations, inspiration, and creativity suggests that these higher level "laws" are actually law statements. There are no new nomological influences at work in these higher levels, but there are conceptual and theoretical innovations that must be introduced in order for us to understand and render tractable the complexity of phenomena occurring above the fundamental level. In a later article, Anderson confirms this reading: "This, then, is the fundamental philosophical insight of twentieth century science: everything we observe emerges from a more primitive substrate, in the precise meaning of the term 'emergent,' which is to say obedient to the laws of the more primitive level, but not conceptually consequent from that level" (Anderson 1995, 2020).

This is a consistent position—a particular type of fundamental physicalism supports the nomological realist claim that only the laws of fundamental physics exist as features of the world, and the conceptualist attitude toward the predictive and organizational regularities that are used outside the fundamental domain supports the attitude that these are to be construed as reflecting artifacts of the representational apparatus.

3. I provide a detailed exegesis of Anderson's article in Humphreys (2015). Here, only the broad outline is given.

A variety of arguments can be brought to bear in support of the conceptual part of this position: the insurmountable computational difficulties involved in many ab initio calculations, symmetry breaking, and the existence of singularities at the $N \rightarrow \infty$ limit are three such reasons, but the conclusion is clearly conceptual in nature: "Surely there are more levels of organization between human ethology and DNA than there are between DNA and quantum electrodynamics, and each level can require a whole new conceptual structure" (Anderson 1972, 396). The new representational apparatus makes predictions and theoretical integration easier, but ontologically, there is nothing new at the emergent level.

A reasonable interpretation of Anderson's position is, therefore, to take it as suggesting that the behavior of all matter is constrained by fundamental laws, but that in order to achieve predictive efficiency the constructionist project requires the introduction of new concepts and law statements at various stages of the project, where a concept is a representation of a property. The mode of representation can vary. Usually representations will be linguistic, but the novel apparatus could include graphical representations.

The importance of Anderson's shift in direction is considerable. In emphasizing reduction, rather than construction, most philosophical discussions of theoretical reduction have been conducted under two assumptions: the assumption that the concepts and linguistic resources of both the reducing and the reduced domains already exist, and the assumption that reduction involves the derivation in principle, rather than the derivation in practice, of the reduced theory. For the constructionist project, we need to abandon both of those assumptions, at least in part. There are many situations, in macroeconomics, in condensed-matter physics, in sociology, and in many other areas, in which a new vocabulary is deliberately introduced in order to provide an effective treatment of the subject matter rather than relying on the vocabulary appropriate to the constituent entities of the entities at hand. Considerable ingenuity is often required to invent an effective framework. With regard to in-principle reduction, any claim about reduction or construction that was not supported by at least some partial success in actually carrying out that reduction or construction is merely an article of faith.

Insightful as Anderson's suggestions are, they are still operating under the constraints of the levels picture. The inability to construct descriptions of nonfundamental$_d$ (and fundamental$_S$) states, properties, or other entities also occurs in diachronic situations; cross-temporal predictions can require a representational apparatus that transcends even the step-by-step simulation methods of weak emergence. If the computational complexity of the simulation process is incompatible with the capabilities of the available computational devices that implement the process, or the state that needs to be predicted is undecidable with respect to the theory governing the initial states, then predictive effectiveness will require the introduction of a new representational apparatus or the injection of measured quantities or observed states into the predictive process. In the following sections, we shall see how this orientation toward synchronic reduction and construction has affected attitudes toward conceptual emergence.

5.2 REDUCTION AND CONSTRUCTION

Reduction can take place between *domains of science,* such as the reduction of sociology to sociobiology; between *theories,* such as thermodynamics and statistical mechanics; between *terms* in theories, such as "water" and "H_2O"; between *objects,* such as social groups and their members; between *natural kinds,* such as hydrogen and a particular atomic structure; and between *properties,* such as being anemic and having a hemoglobin deficiency.[4] In the special case where the reducing domain is physics, reductionism becomes a version of physicalism, whether fundamental or nonfundamental. Within the categories of property reduction and object reduction, we have a distinction between type–type reduction, where the reduction is between types of property or types of object, and token–token reduction, where the reduction relation holds between instances of properties or between specific objects. Token–token reduction can hold even when type–type reduction fails,

4. I am not claiming that these reductions are correct; rather, they are examples of the kinds of reductions that scientists have attempted.

a feature that is closely associated with multiple realizability in the property case.

Emergence is usually taken to be incompatible with reduction in the sense that A's being irreducible to B is a necessary condition for A to emerge from B. Irreducibility cannot be a sufficient condition for emergence because two different types of fundamental entities are irreducible to one another, but there is no relation of emergence between them. Each category in our taxonomy of emergence has its counterpart type of irreducibility—ontological irreducibility, inferential irreducibility, and conceptual irreducibility—and to be specific, the correct claim is that ontological emergence is incompatible with ontological reduction and similarly for the other two categories. On the other hand, ontological reducibility is compatible with inferential emergence and with conceptual emergence. When considering theoretical reduction, the truth of the theories involved is not a requirement for a successful theoretical reduction. Thus, if we do not insist on the truth of the theories between which a theoretical reduction relation holds or on the conceptual correctness of a representation, then the ontological emergence of type B entities from type A entities is compatible with both the theoretical reduction of a theory about Bs from a theory about As and the conceptual reduction of a representation of B to a representation of A. Other compatibilities are possible; the point is to be clear about which part of the taxonomy we are dealing with.

We must also distinguish between ontological reduction and ontological eliminativism. For ontological eliminativism to be true about some entity E, we must be in the situation where E was once believed to exist, but subsequent evidence has established that this belief is false, and whatever evidence was once taken to support E's existence can be accounted for in terms of other entities. A famous example of eliminativism is Henri Bergson's élan vital, the force supposedly responsible for evolution, the functions of which are now known to be performed by the objects of molecular biology, the processes of self-organization, and natural selection. In contrast, when ontological reduction occurs, the reduced entities are identified with the reducing entities, and so exist just as much as do the reducing entities, albeit usually under a different mode of description. Assuming that it is appropriate to claim that my bicycle can be reduced to a properly structured collection of mechanical parts, my bicycle has not been eliminated; it has just been shown to be identical with that structured collection. Continued reference to the reduced entity is a matter of convenience.[5]

For good or ill, a considerable part of the philosophical literature on reduction has been a reaction to the account of reduction by Ernest Nagel. Within Nagel's account, one theory is irreducible to another if the law statements of a higher level theory cannot be deduced from those of a more fundamental theory by employing "bridge laws"—the term "laws" here is potentially misleading—that connect the two levels. Thus, if a theory T′ is Nagel-irreducible to a theory T, the combination of the law statements of T and the use of bridge laws to translate the vocabulary of T′ into the vocabulary of T is insufficient to allow every statement of T′ to be derived, in principle, from statements of T. Any law statements that contain only observational terms are theoretically neutral, and so do not need to be reduced. To carry this out in practice is usually enormously complicated and often impractical. Reduction and emergence are connected in Nagel's approach because he took inferential emergence to consist in the underivability of sentences about higher level properties from sentences about lower level properties (Nagel 1961, 367). His approach can also accommodate nomological emergence when law statements are the objects of study.

A primary constraint on Nagel reduction is the computational complexity of carrying out the required derivations. Frequently, the computational resources that would be required to accomplish the calculations for an in practice ab initio reduction are so far beyond feasible capabilities that the calculations could never be completed on a standard computing device. An example is the calculation of the exact properties of a macroscopic piece of stainless steel from the quantum states of its constituent atoms. A second reason for the constraint is that with continuous-state variables, the errors introduced by finite approximations often need an

5. There are other types of eliminative reduction, such as when a vague classification is refined to become a more precise taxonomy. One example of this is the replacement of the concept of shell shock by the concept of posttraumatic stress disorder. I shall not discuss this form of eliminative reduction here.

infeasible amount of computing power to reduce to acceptable levels. One caveat is needed when considering such constraints: the degrees of computational complexity involved in the calculations are defined relative to a particular class of computing devices—those equivalent to Turing computability. Were quantum computing or some other radically different form of architecture to become available on a large scale, certain intractable problems would become feasible. These considerations about degrees of computational complexity may seem relevant only to the sweaty factories of applied science. Exposure to the realities of theoretical reduction will quickly temper that view.

Nagel reduction comes in two kinds: homogeneous and inhomogeneous. Homogeneous reduction concerns cases in which the vocabularies of the two theories are the same, whereas in the inhomogeneous case, the theory to be reduced has terms that are not present in the reducing theory. Conceptual emergence primarily occurs when inhomogeneous Nagel reduction fails, and it is with this type of reduction that we shall be concerned.

Inhomogeneous reduction rests on two ideas. The first is that when we are faced with the task of reducing one theory to another, and the theories use different vocabularies, a successful reduction has to tell us how those vocabularies are related to one another. Nagel suggested that this should be done by constructing bridge laws. One, admittedly contentious, example is a bridge law between temperature (as used in macroscopic physics, chemistry, biology, and elsewhere) and the mean kinetic energy of molecules: If *x* is an ideal gas in thermodynamical equilibrium, then *x* has temperature T if and only if the molecules of *x* have average kinetic energy E.[6] In cases where bridge laws cannot be constructed for the different vocabularies of the two theories, and as a result some sentences of T′ containing predicates specific to T′ cannot be derived from sentences entirely in the vocabulary of T, we have a situation that is a candidate for conceptual emergence.

A major advantage of these bridge laws, if they can be formulated in the manner just given, where an equivalence is provided between the

6. For a detailed discussion of this case, see Sklar (1993).

higher level term and the lower level term, is that this allows the higher level term to be eliminated from the theoretical language. This linguistic elimination does not entail ontological eliminativism as described above, even when the respective theories are true. It only, in conjunction with the appropriate derivations, gives us theoretical reduction. Temperature still exists, but it has now been identified with mean molecular kinetic energy.

5.2.1 A Turn to Ontology

The conceptual apparatus we have discussed thus far reflects the synchronic or atemporal contexts that occur with theoretical reduction and determination relations between levels such as supervenience, with the primary concern being a re-representation of the same set of entities within the ontologically preferred level. A central question regarding Nagel reduction is whether the bridge laws are necessarily true definitions rather than contingent identity claims or regularities. Recalling Broad's (1925, 77–78) account of transordinal laws, we see that transordinal laws are regularities that can be known only a posteriori and cannot be deduced, even in principle, from laws in other domains. So, within Broad's approach, this rules out a priori conceptual or deductive connections between the emergent domain and other domains, whether based on definitions or on some other epistemic source such as intuition.

The inability to derive some claims about phenomena in domain B from a theory about domain A is evidence that something in domain B which is essentially new with respect to A is being described, and that A is deductively incomplete with respect to the B phenomenon. The theoretical and conceptual resources of the theory about domain A do not, implicitly or explicitly, contain the required content about this new phenomenon.

Consider how this kind of theoretical reduction works for the Checkers World of section 1.4. Within this model, levels serve no useful purpose and a diachronic approach is more suitable. The blue particles have diachronically emerged from the white particles, and to describe the behavior of the blue particles requires concepts that do not apply to the white particles, such as "moves backward" and "annihilates red particles." But these concepts do not involve a different "level"; the first is geometric, whereas the concept of a red particle can be co-defined with the concept of a blue particle. In addition, the law that connects the juxtaposition of a white particle with the topmost boundary of the board and the subsequent appearance of a blue particle can be learned only by induction. There is nothing in the theoretical apparatus governing the white and black particles from which that law can be derived.

In reductive projects, we start with the higher level theoretical apparatus and often do not at that point know what the nature of the lower level reducing entities are. The reductivist project requires us to discover them and to provide an appropriate conceptual apparatus that represents them in a way suitable for the reduction project. With construction, we start with the lower level concepts, and sometimes do not know what the higher level features are or should be, and an appropriate vocabulary must be developed for the construction project to be carried out. The bridge laws or the transordinal laws in both cases, therefore, sometimes require the invention of new concepts and not just the discovery of connections between existing concepts.

An alternative to Nagel reduction that is compatible with ontological emergence, although not with inferential and conceptual emergence, and which makes more of a concession to applications is Kemeny-Oppenheim reduction. A theory T_1 is Kemeny-Oppenheim reducible to theory T_2 if and only if all observational data explainable or predictable by T_1 is explainable or predictable by T_2. Any theory that is Nagel reducible is Kemeny-Oppenheim reducible, but not vice versa, because Kemeny-Oppenheim reducibility does not require the existence of bridge laws. A principle of ontological priority can then be based on Kemeny-Oppenheim irreducibility: If the existence or properties of an entity A can be explained in terms of another entity or collection of entities B, but not vice versa, then B has ontological priority over A. This is a sufficient condition for ontological priority, but it is not necessary because supervenience relations often do not permit explanations. However, it is generally taken that if A supervenes on B, but not vice versa, then B has ontological priority.

5.3 PHILOSOPHICAL COUNTERPARTS TO CONSTRUCTIONISM

If one were to choose a principle that captured the spirit of twentieth-century analytic philosophy, Russell's maxim of logical synthesis would be a prime candidate: "The supreme maxim in scientific philosophizing is this: Wherever possible, logical constructions are to be substituted for inferred entities" (Russell 1914, section VI).

Russell was writing of constructions in terms of sense-data (which were, for him, physical entities), but the method he endorsed generalizes when logical constructions are supplemented with mathematical constructions.[7] In what follows I shall construe Russell's maxim in that broader sense. Russell's advice suggests that only a limited use of inference to the best explanation is appropriate. Moreover, in the case of constructionism as considered in this chapter, we are not concerned with inferences from the realm of the observable to the realm of the unobservable. This is because we have already accepted the existence of the more fundamental entities and wish to construct the rest of the unobservable realm on that basis. Our primary concern is how the fundamental and nonfundamental entities are related. Because of this, we are not constrained by the epistemic foundations of traditional empiricism.

Russell also subscribed to a form of Occam's Razor, the adage that "Plurality is not to be assumed without necessity." There are different degrees of ascetism involved in using Occamist principles, but within the constructionist tradition the idea is that if we can capture, without residue, the content of a given domain that has an apparently autonomous ontology (for example, the domain of mental states) by using a minimal preferred ontology (such as physical states) and logico-mathematical constructions from it, then we can and should give up any commitment to the surplus ontology in order to avoid incorrectly asserting the existence of a separate domain of entities. Thus, we have a second maxim of Russell's: "That is the advantage of Occam's Razor, that

7. Since the logicist program of reconstructing mathematics using logic failed to fully achieve its goals, the generalization is necessary.

it diminishes your risk of error" (Russell 1918; also Russell 1956, 280). Sound advice, but razors, especially of the fourteenth-century variety, can be dangerous. By keeping our ontological commitments minimal, Occam's Razor helps prevent us from asserting the existence of entities that do not exist, but that very minimalism can lead to another kind of error—that of refusing to admit into our ontology things that do in fact exist. These are philosophical analogues of the statistical concepts of Type I and Type II errors, respectively, and the austerity of the generative atomism tradition can lead to errors of the second kind when ontological emergence is present. As the caveat "wherever possible" in Russell's supreme maxim indicates, circumstances sometimes require enlarging our ontology and recognizing that logical constructions are insufficient. In cases of transformational emergence, the emergent entities cannot be synchronically constructed from prior entities, and inferences about the properties of the novel entities must be made, often using inductive methods.

Yet acknowledging that ontologically there is only one domain, rather than two or more, leaves the constructionist with Anderson's problem: Do we need to introduce a new conceptual apparatus in order to facilitate prediction and description to augment the construction process? Nomologically holistic properties often require conceptual novelty. Recall that a nomologically holistic property is one that could not be possessed by the constituents of the object possessing the property—it is nomologically impossible for this to occur. Although the point is pitched in terms of properties and their corresponding predicates, it carries over to law statements. As an example of the role that systemic properties can play in generating novel regularities that are candidates for nomological emergence, Berlekamp, Conway, and Guy (2004) proved that a cellular automaton programmed with the rules for the Game of Life can serve as a universal Turing machine. Because the principal computational roles are played by structured sets of cells called gliders (see section 4.2.7), and not by individual cells, the domain of cell states is not the domain within which the computations are taking place. Thus, as Russ Abbott (2006) has noted, because the original programmed rules for the cellular automaton do not include rules for glider interactions, the regularities exhibited by gliders provide an opportunity for nomological emergence. Furthermore, the concept of "carrying out addition" has no meaningful content at the level of the cells; rather, it is only at the level of the emergent forms such as gliders, and so new representational content is needed to capture the Turing machine computations.

Conceptual innovation does not always involve new terminology. Fritz Rohrlich (1997, S347) has noted that a term such as "gold" can occur within two different theoretical domains and have quite different meanings within each domain.[8] This fact has three specifically philosophical consequences. First, it casts doubt on the general effectiveness of the examples used by Hilary Putnam (1962) to argue that there could be no specifically observational predicates such as "red" because "red" could also be applied to unobservable entities. In the case of gold, it has been empirically determined that at the nanoparticle level, gold has different properties than at the macroscopic level—at the macro level, gold is chemically inert, whereas at the nano level, it acts as a catalyst to convert carbon monoxide and oxygen into carbon dioxide. (Realist claims that such things as nano-level gold exist can be justified by the dilution argument found in Humphreys 2004, chap. 2.) If we use the causal criterion for property identity (Shoemaker 1980), the predicate "gold" used within the molecular domain will pick out a different property than will the predicate "gold" used within the macroscopic domain. So, these are the same predicate only in the trivial syntactic sense of being polysemes (or perhaps homonyms).

Second, since "gold" picks out different properties in the two domains, this also calls into question the appropriateness of Nagelian homogeneous reduction when applied in certain cases, because Nagelian reduction uses only syntactic criteria to identify terms across the domains. What is at issue are the various properties picked out by the term "gold" in the different domains, and we cannot identity these using only local syntax. Third, this fact suggests that the famous identity "water = H_2O" may well be true for the term "water" when it refers to water molecules, but false when its homograph refers to the macroscopic

8. Conceptual emergence is also suggested by Steels (1994, 89).

liquid. A fortiori, the identity is not necessarily true with the universal scope often attributed to it.[9]

5.4 FUNCTIONAL REDUCTION

A different approach to reduction, which I shall call "functional reduction," has been suggested by a number of authors as a more powerful alternative to Nagel reduction.[10] I shall primarily discuss the version given in Kim (1999), since it is the most fully developed. According to these authors, if we approach reduction correctly, most properties turn out to be reducible. Then, if reducibility precludes emergence, there will be only a few properties that cannot be functionally reduced, and hence are candidates for emergence. Among these will be features such as qualia and consciousness.[11]

For Kim, there are three core ideas associated with emergent properties: novel causal powers, unpredictability, and unexplainability in terms of the system's microstructural properties. The total microstructural property of a system is the property determined by a description in terms of (a) the basic particles that constitute the system, (b) all the intrinsic properties of those particles, and (c) the relations (not necessarily just spatial relations) that configure the particles into a structure.[12] Because the total microstructural property belongs to the entire system, it is a macro property. The core idea of functional reduction is to find a realizer in domain B of a given property in domain A, with a

9. See Johnston (1997) for different arguments to the same conclusion.
10. The approach has also been explored in Levine (1993), McLaughlin (1997), Kim (1997, 1999, 2005), and others. McLaughlin, for example, writes "properties that are susceptible to functional analysis in physical and topic-neutral terms are reducible" (1997, 36) and "The members of the British Emergentist tradition apparently failed to appreciate, however, that dispositions and capacities that are functionally analyzable are ipso facto reducible" (36). For objections to Kim's version of the approach, see Anthony (1999).
11. "Functionalization of a property is both necessary and sufficient for reduction (sufficient at least as a first conceptual step, the rest being scientific research).... [I]t explains why reducible properties are predictable and explainable" (Kim 1997, 13).
12. Kim (1999, 6–7).

view to using the predictive power of domain B and its ability to explain phenomena in terms of the microstructural properties of entities, so that the putative emergent properties are redundant.

More specifically, there are three steps that make up the jointly sufficient and individually necessary conditions for reducing one property E to a collection B of other properties:

1. *Functionalize E.* That is, construe E as a property defined by its causal or nomic relations to properties in the reduction base B (either directly or through further reductions). This provides a conceptual connection between the two domains through this functional definition:

> *Definition* Having E = Having some property P in B such that (i) P is typically caused to be instantiated by $C_1, \ldots, C_n$ and (ii) P typically causes $F_1, \ldots, F_m$ to be instantiated, where each C_i and F_j is in B. Either (i) or (ii) may be vacuously satisfied and P is the "realizer" of E.[13] Thus, if we know that such an instance of P is in place, we know without further investigation that an instance of E is present.

An important role is played in this process by the Causal Inheritance Principle, which asserts: If a functional property E is instantiated on a given occasion in virtue of one of its realizers Q being instantiated, then the causal powers of this instance of E are identical with the causal powers of this instance of Q. This principle allows us to account for why the dependent property is attributed the causal powers that are ultimately due to the foundational features. The principle is needed because functionalizing the dependent property automatically requires that it have causal powers, and so is considered to be real according to Alexander's Dictum.[14] Finally, the original account assumes that we have a solution to the problem of differentiating levels.

13. Kim claims that this kind of approach can be found in Armstrong (1972) and Lewis (1966), but that neither Armstrong's nor Lewis's position is reductionist, and that most functionalist positions are also not reductive.

14. Alexander's Dictum, named after the British emergentist Samuel Alexander, asserts that a property is real if and only if it has causal powers. Alexander, in writing "the doctrine that mind is an epiphenomenon . . . is not simply to be rejected because it

2. *Find realizers of E in B.* This step, which requires scientific investigation, is crucial to successfully carrying out the reduction.

3. *Find a B-level theory that explains how realizers of E perform the causal task constitutive of E.* For example, a gene is functionally characterized as "the mechanism in a biological organism causally responsible for the transmission of heritable characteristics from parents to offspring."[15] In humans and other species, genes are realized by DNA, RNA, and associated entities, and a systematic theory within molecular biology explains how those realizers work. Kim noted that functional reduction will usually be a piecemeal business, and we shall need to be satisfied with things like the reduction of human pain to human neurophysiology, rather than a reduction of pain in general, because of the heterogeneity of the realizers.

Despite the considerable importance of the functionalization approach in functional reduction, the key difference between this account and the earlier but largely abandoned approach of Nagel reduction lies in the nature of the connections between the two domains. Kim's main objection to Nagel reduction is that the Nagelian bridge laws essentially extend the linguistic or conceptual resources of the reduction base, and the bridge laws themselves require explanation, making Nagel reduction irrelevant to the problem of the explanatory gap between the physical and the mental. In Nagel's approach, the bridge laws that connect the domains are contingent and must be discovered by empirical or theoretical inquiry in the sciences.

In contrast, one of the characteristic features of functional reduction is that those bridge laws are replaced by conceptual connections between E and the properties in B that are identified during the

supposes something to exist in nature which has nothing to do, no purposes to serve, a species of *noblesse* which depends on the work of its inferiors, but is kept for show and might as well, and undoubtedly would in time be abolished" (1920, 8) somewhat unnervingly improvises on Russell's infamous quote regarding causation: "The law of causality, I believe, like much that passes muster among philosophers, is a relic of a bygone age, surviving, like the monarchy, only because it is erroneously supposed to do no harm" (Russell 1912, 1).

15. See Kim (1998, 25).

functionalization process. It is this conceptual connection that allows us to move from the properties in the preferred domain to the properties in the dependent domain, and the functional definition allows us to keep the scientific explanation entirely within the realm of the preferred properties B. This is why only inductive predictions are possible with Nagelian reduction, whereas theoretical predictions based entirely on B-domain laws are possible when functional reduction has been completed. The functional-reduction approach to emergence (in the rare cases that it exists) is thus squarely in the conceptual tradition we have discussed, where novel logically independent concepts must be introduced to represent the emergent entities.[16]

The few properties such as consciousness that Kim considers as candidates for being emergent are not predictable on the basis of theory. They can be inductively predicted if the repeated appearance of a novel macro property whenever a given microstructural property is present allows us to discover a regularity and use it to predict the next case of that macro property in a way that could not have been done before the data was gathered. In contrast, the physical properties of the system are predictable ab initio at the micro level in virtue of the theories that apply to that level. A failure to predict an emergent property on the basis of the theoretical statements that apply to the micro level may be, as in the case of qualia, because their intrinsic qualities are beyond the reach of any inter-subjective representational apparatus.[17] This difference between the nomic predictability within the micro and macro domains results in the characteristic epistemic irreducibility of transordinal laws or associations.

The most direct way in which functional reduction can fail is in cases where the higher level property cannot be functionalized. This does

16. "[W]hat enables the ascent from the reduction base to higher properties is the conceptual connections generated by the functionalization of higher properties. This is in sharp contrast to Nagelian reduction with bridge laws taken as auxiliary premises" (Kim 1999, 14).

17. Kim denies that the practical difficulties of prediction are relevant: "But obviously ease and simplicity of calculation as such is of no relevance here; predictability is not lost or diminished if calculationally complex mathematical/logical procedures must be used" (Kim 1999, 7). This very strong dispositional sense of "predictable" ensures that we will lose contact with many examples that motivate the inferential emergence position.

not mean that the higher level property would be causally isolated—a "causal dangler"—because it could still enter into causal relations as an emergent property having its own, distinctive causal powers. That is not an objection to the approach; it simply means that the class of irreducible properties depends crucially on how broadly applicable is the functionalization approach, and it is a contingent fact about the world how many properties exist that are resistant to the functionalization procedure. For example, it is not possible to functionalize the role of the blue particles in Checkers World because the only candidates for the realizers are the black or white particles, and these do not have the right causal features.

Kim's view is that there are no emergent properties in his sense within the physical realm: "For there may not be any emergent properties, all properties being physical properties or else functionalizable and therefore reducible to physical properties" (Kim 1999, 18), reinforced by the later claim (20–21) that "A characteristically emergentist doctrine makes its appearance in the idea that some of the properties of these complex systems, though physically grounded, are nonphysical." Given this view that emergence is rare, the default position must be that functionalization is iterative. The realizer property at any given step can itself be functionalized, and the process repeated until we reach fundamental properties that themselves have no realizers.[18] The conceptual process used in the functionalization step is synchronic (or atemporal), and so it is compatible with distinctively diachronic forms of emergence. It is also compatible with weak emergence, it allows scope for transformational and fusion emergence, and in the synchronic realm, it allows room for conceptual emergence.

18. On the assumption that there is no infinite regress of such properties.

6

Philosophical Topics Related to Emergence

6.1 PHYSICALISM

Unlike mathematicians and logicians, philosophers do not have firm control of their technical vocabularies. A mathematician who employs a deviant use of the term "continuously differentiable" will be quietly labeled incompetent and ignored by the rest of the profession. In contrast, the terms used by philosophers frequently do not have sharp conditions of application, they are often not provided with explicit definitions, and a philosopher who employs a variant use of a term might well be hailed as doing something creative and informative. This, unfortunately, is the situation with respect to physicalism, which even with significant efforts over the last two decades, lacks a standard definition.[1]

Even so, we can distinguish between two broad kinds of physicalism—*fundamental physicalism* and *nonfundamental physicalism*. Fundamental physicalism, as it is usually construed, always takes a synchronic form, and so no subscript is necessary. It identifies certain physical entities and properties as basic, these being whatever particles, forces, spaces, fields, or other entities are currently considered by physics, or would be so considered at the limit of foreseeable inquiry, to constitute the ultimate physical ontology. It then has to account for the remaining properties and objects of physics, as well as those of chemistry, biology, and other ontologically acceptable sciences. When explicit

1. For persuasive arguments that there is no true, nonvacuous characterization of physicalism, see Crane and Mellor (1990).

reduction or construction appears infeasible, fundamental physicalism frequently appeals to a determination relation or relations. The determination relation can be one of supervenience (of which more in the next section), of realization, of determinates and determinables, of abstraction, of second-order property formation, of grounding, or one of any number of similar determination relations. The essential feature of all these relations is that once the entities in the physical domain are present, the properties in all other domains are thereby determined. Fix the brain states, and you have the psychological states (supervenience); fix the computer state, and you have a functional equivalent to a belief (realization); fix the color of burnt sienna (the determinate), and you have the color brown (the determinable); fix the rotational symmetries, and you have a group (abstraction).

The apparatus of determination is largely logical in form. In supervenience relations, for example, Boolean or set-theoretic combinations of basic properties form the subvenient base, with logical, metaphysical, and sometimes nomological necessitation relations providing the appropriate determination relation. Although there is a spectrum of positions about the ontological status of the dependent properties, ranging from eliminativism to a robust realism, for fundamental physicalism to play its foundational role any entities that are not in the fundamental domain must have a secondary ontological status.

This is a view that has proved to be attractive to many modern philosophers: the world is nothing but fundamental physical objects and fundamental physical properties, where the "nothing but" is given substance in terms of the chosen determination relation.[2] Among those physical objects and properties are certain things that are fundamental, in the sense that they synchronically determine all others. One task of the philosopher is to provide a representational apparatus that captures the bare structure of this physical world and to assess what, if

2. The possibility of ontologically distinct domains of inquiry is incompatible with some unity of science positions, particularly those that require a unified ontological basis for science. It is, in contrast, in principle compatible with a unity of method view as those areas of emergence that are unified by a set of methods drawn from complexity theory indicate.

any, ontological commitments follow from this apparatus beyond those already contained within the realm of physical objects and properties.

As presented, fundamental physicalism in the sense just described is compatible both with inferential emergence and with conceptual emergence. It is incompatible with synchronic ontological emergence if existence is confined to the fundamental domain, but as we saw in chapter 2, it is compatible with some versions of diachronic ontological emergence.

A more liberal version of physicalism—nonfundamental physicalism—takes as acceptable a wider domain of entities, which is usually identified with the combined domains of physics, chemistry, and perhaps some parts of biology. The boundaries of the acceptable domain vary widely; for some, the subject matters of geology, meteorology, metallurgy, astrophysics, cosmology, and so on are included even if they are irreducible to the domains of the three areas just mentioned. But always, the intentional entities of the psychological and social sciences are excluded from the domain of the physical.

Nonfundamental physicalism is compatible with all three kinds of emergence, but because much of the contemporary philosophical literature on emergence has focused on the division between the nonfundamentally physical and the mental, and has thereby confined the possibility of emergent phenomena to at most a few aspects of the mental domain, such as qualia and consciousness, it has for the most part failed to address the rich set of emergent features that can occur within the nonfundamental physical domain.[3] The ontological position taken in this book is one of nonfundamental physicalism, but I leave open the possibility that there are ontologically emergent phenomena falling outside the domain of the broadly physical, especially in the social sciences.

6.1.1 Motivations for Physicalism

There are different motivations for physicalism. One is a methodological desire to have in our ontology only things whose existence and properties can be established by scientific methods. To many physicalists, our

3. The philosophy of science literature in contrast does contain systematic discussions of physical emergence.

inability to gain third-person access to most mental entities is a powerful reason to adopt at least nonfundamental physicalism. Both varieties of physicalism exclude from the realm of the physical specifically phenomenal features such as the subjective content of color perceptions, pains, laughter, and so on. Their subjective nature is taken to be incompatible with the methodological motivation for physicalism, and in cases such as color, where physical correlates such as wavelengths of electromagnetic radiation have been identified, the associated qualia are distinct from those physical correlates.[4] Although I shall not pursue this issue, the first-person access to qualia should not by itself rule out the possibility of a scientific treatment of these phenomena, nor exclude them as ontologically emergent. Unless we are sympathetic to some variety of solipsism, we do believe there is evidence for regularities in human experiences of qualia. Tests for color blindness are not the result of conformity to a common language; it does not matter what we call the colored dots on an Ishihara colorblindness plate—what counts is whether or not a spatial pattern can be observed among the intrinsically colored dots.

This methodological motivation is similar to the motivation for naturalist positions. As with physicalism, there is anything but unanimity about what naturalism is. One common variant contrasts the natural with the supernatural, and uses that contrast as the basis for preferring the former over the latter on the grounds that the former is amenable to scientific methods whereas the latter is not. Anything supernatural is to be excluded, and that means, in addition to theological and spiritual entities, specifically metaphysical entities. Naturalism is compatible with nonfundamental physicalism, since biological entities are perfectly respectable natural entities and by common assent they are not fundamental physical entities. Some earlier versions of emergentism such as those of Alexander (1920) had sympathies with supernaturalism, and there are contemporary approaches to emergence that maintain those

4. Unlike mental entities such as beliefs, tokens of experienced qualia ordinarily have identifiable spatial and temporal locations, making an identification with physical entities at least possible. A claim that this can be done by identifying qualia with either physical and chemical features or with functional properties can be found in, for example, Tye (1989, 133).

sympathies, such as Clayton (2004) and Murphy (2006), which have taken emergence to allow the possibility of supernatural entities such as the human soul. The position taken here is that all the clear current examples of emergence come from scientific investigations, and there is no evidence for supernatural emergence. Although there could be scientific evidence for what are currently considered to be supernatural phenomena, such as ghosts, no persuasive evidence has been produced, and souls are even more elusive than ghosts.

The arguments given in this book are sympathetic to naturalism, but any version of naturalism that denies by fiat the existence of abstract entities and the contents of conscious experience is too constrained. One does not have to be a dualist to regard the contents of one's own conscious experience as real, and if those experiences are emergent, at least one type of emergent feature is first-person accessible only. And if it turns out that abstract objects exist, their existence should not be denied on methodological grounds alone.

There can be a second motivation for physicalism. It is that physics is the most general science, and physics is important not just because it studies matter in the most fundamental domain but also because at least some of the properties it studies, such as energy, occur in all domains that are considered to be naturalistically legitimate, and therefore these properties form a privileged part of our ontology. No other sciences have this generality because only physics has access to the most fundamental level.[5] This kind of motivation cannot justify nonfundamental physicalism unless it can be shown that all genuine properties are domain neutral, which is implausible.

What cannot serve as a motivation for physicalism, although it seems that for many this has been a motivation, is the argument that mental entities, vitalism, and so forth must be rejected as real because they are too strange. This position is immediately vulnerable to a tu quoque reply: the entities of current fundamental physics and astrophysics

5. It is possible that in the future some fundamental science will be developed that falls outside the domain of physics, but this currently seems unlikely in that the imperial sweep of physics is extremely powerful and tends to subsume even immaterial ontologies having wide application, such as information.

are even weirder than the usual mental candidates, and limit physics is not likely to get better in that respect. The ontological strangeness of contemporary physics needs no further comment, and there is no clear reason why digital physics, which holds that information is the basis of physics, and thus abstracts from the physical realization, should be preferred to mental entities. Probabilities, whether construed as degrees of rational belief or viewed more objectively, play a fundamental role in physics and are as mysterious as are beliefs—even energy is not above suspicion.

6.1.2 Limit Physics

Discussions of physicalism sometimes invoke the idea of limit physics. Like emergentism, physicalism, whether limit physicalism or some other type, should be a contingent thesis supported or undermined by empirical evidence. This feature strongly suggests that any discussion of what counts as falling within the domain of physics must confine itself to the scope of contemporary physics. This is because any appeal to limit science, where limit science is the science that would be in place if scientific inquiry employed its methods to arrive at a final, unrevisable set of theories, lacks knowable content. We know what the limit is of the mathematical sequence $1/2, 1/4, 1/8, \ldots, 1/2^n, \ldots$ because we can deduce it a priori, but we do not know what the complete set of truths of limit science will be, in part because discontinuous changes in concepts that require creativity occur as physical theories change, and in part because new evidence will be discovered that is unpredictable from current theories.[6] If we did have that knowledge, we should already be at the limit, and as a result, we should already have evidence for scientific realism and have solved the problem of the pessimistic induction.[7] That argument presupposes that science is a truth-seeking activity, where truth is construed in the correspondence account. Although some versions of

6. This argument can be found in Jardine (1980).

7. The pessimistic induction is the argument that the history of science exhibits periodic overthrows of well-confirmed theories, and that there is no reason to think that this historical pattern will change in the future.

limit science take an instrumentalist or pragmatist attitude toward the limit theories, because physicalism is an ontological claim, those versions are not relevant here.

Any claim that we have access to the content of limit physics is implausible—imagine the physicists of 1687 attempting to predict what twenty-first-century physics would be like. Although Newton would have been intrigued by and, given some time, would have been able to understand, current physics had he been presented with it, he had no means of anticipating in his own era the existence of such developments as quantum field theory, because the experimental evidence that led to its formulation was not available to him. We are in an even worse position with respect to physics hundreds of thousands of years in the future, and there are reasons to think that humans will be unable to understand the detailed content of much of future physical theories. Finally, the appeal to a limit does not help matters, because there is no closed list of properties that count as physical; if we characterize something as physical just in case it is part of the domain of physics, this leaves open almost anything to count as physical. Physicists currently investigate complex systems, biophysics, astrophysics, and other areas, none of which are obviously "physics" in the sense that fundamental physicalists intend. In fact, the very idea of limit science may well not have any clear sense unless we fix what can count as a possible observation. If, as I have suggested elsewhere, the realm of the observable continually expands, there may be no limit to the kinds of observations it is possible to make.[8] Even as a bare existence claim, the concept of limit science is suspect because we have no proof that the limit exists. There could be an infinite sequence of replacement theories, each incommensurable in roughly the way Kuhn (1962) claimed.

Finally, "limit physics" does not mean "limit fundamental physics," and the tendency to think that, at the limit, physics will have been completely reduced to fundamental physics must be therefore be resisted. To do otherwise is to beg the question against the existence of emergent phenomena within physics. Limit science should also be distinguished from complete science. A scientific theory T is complete if every true

8. See Humphreys (2004).

proposition about the subject matter of the theory is a logical consequence of T, and the theories of limit science need not be complete in that sense. Contemporary science uses mathematical theories that are incomplete, and this situation may well persist at the limit of science. In addition, there are physical systems models of which are known to have undecidable states. These issues do not affect ontological emergence, but they are central to claims about inferential and conceptual emergence.

6.2 EMERGENCE AS SUPERVENIENCE

One way that has been suggested of capturing emergence, and which is motivated by physicalism, is based on supervenience relations.[9] Despite internal differences in the accounts, a common element to these approaches is the conclusion that emergent phenomena are rare, and are confined to qualia and associated aspects of conscious mental states. Because we have seen examples of ontological emergence within the physical domain, and that ontological emergence, although unusual, is not as rare a phenomenon as these approaches claim, it is important to understand why these supervenience accounts are too weak to provide an understanding of ontological emergence, even though they can provide an apparatus for representing conceptual emergence. I shall give a formal definition of supervenience later, but for the moment, the general idea is that a property B supervenes upon a set of properties A just in case B's presence is fixed by the presence of the members of A, and it is fixed not accidentally, but necessarily. Similar approaches can be used for objects or concepts instead of properties.

9. See, for example, van Cleve (1990), Teller (1992, 142), McLaughlin (1997), Kim (1992), Chalmers (1996), and Yates (2009). Butterfield (2011a) takes supervenience to be a form of reduction that allows infinitely long definitions. In some areas of philosophy, supervenience relations have been superseded by grounding relations. I have cast the discussion in terms of the former, in part, because the current emergence literature does not extensively employ the latter. Most of what is said here can be reconstrued in terms of other relations of ontological dependence.

The reasons that supervenience relations are brought into play vary, and the most common do not directly appeal to emergence. Among the arguments in favor of using supervenience relations within various domains of science, three dominate. The first is the failure of Nagel reduction. Because supervenience is a less demanding condition than Nagel reduction, substituting the former for the latter makes physicalism easier to defend. As we have seen, difficulties for Nagel reduction can arise either because of syntactic and deductive complexities or because some concepts needed for the derivation of the higher level are lacking in the lower domain and no satisfactory bridge principles can be constructed. So, in place of an explicit deductive relationship between sentences, the concepts of the higher level domain are taken to supervene upon those of the lower domain. This gives the higher level concepts a dependent, subordinate status in line with the inclinations of a non-eliminativist orientation.

A second argument for adopting supervenience rests on the multiple realizability of properties in certain realms, especially in the realm of the mental. It is claimed (usually without much substantive evidence) that a property such as being in pain can be realized by quite different physical systems in different organisms. Thus, although it is implausible that the concept of being in pain can be reduced to a unitary neurophysiological concept, especially by means of an explicit definition, being in pain can still be construed as being fixed by physical properties in the sense that there can be no change in the mental property of being in pain without an attendant change in some underlying physical property. Multiple realizability opens up the possibility of a certain type of autonomy for the higher level because of the lack of dependence of the higher level on the specifics of the lower level implementation.

A third argument for supervenience rests on the mismatch of natural kinds that appear in laws at different levels; it is possible but unlikely that the properties used in psychology to describe emotions will line up exactly with the properties used in affective neuroscience, and hence a Nagelian reduction cannot be carried out. Even a more fine-grained functional reduction will be difficult to achieve in a way that preserves the unity of the higher level concept. This mismatch often results from the multiple realizability of the higher level property, but could occur even with single realizability.[10]

One preliminary caveat: when supervenience relations are relations between properties, they are neither synchronic nor diachronic because properties are atemporal. Nevertheless, it is common to take them as attempts to represent synchronic cases of emergence, in that within standard definitions of supervenience, any instance of the emergent property occurs simultaneously with an instance of the subvenient property.

6.2.1 Nomological Supervenience

Because it is one of the clearest accounts available within the supervenience approach, I shall take as the focus of my discussion the analysis of emergence due to Brian McLaughlin. McLaughlin's account, as is usual with supervenience accounts, emphasizes mental properties and is motivated by two desires. The first is to leave as an open question whether mental properties are emergent. The second is to ensure that reductive materialism is false if emergentism is true.

McLaughlin's definition is: "If P is a property of w, then P is emergent if and only if (1) P supervenes with nomological necessity, but not with logical necessity, on properties the parts of w have taken separately or in other combinations; and (2) some of the supervenience principles linking properties of the parts of w with w's having P are fundamental laws" (McLaughlin 1997, 39). For McLaughlin, a fundamental law is one that is not metaphysically necessitated by any other laws, with or without initial conditions.[11]

A few things to note about this definition: It appears to assume that the parts are proper parts, and hence that the possessor of the emergent property is composite. It thus does not cover the Checkers World

10. The *locus classicus* for this argument is Fodor (1974).

11. An important aspect of McLaughlin's article is his canvassing of a functionally driven approach to reduction that has important similarities to Kim's account of reduction. I shall not discuss that aspect here. I also note that McLaughlin does not advocate the account of emergence he has constructed.

example or fusion as cases of emergence. In doing so, it follows some of the British Emergentist approaches both in referring to properties of the components in other configurations and in requiring that the nomological link between the levels is fundamental, but predictability has been replaced by an appeal to supervenience. In McLaughlin's approach, the laws of nature are not a fixed part of the supervenience base, but are allowed to vary across possible worlds. Finally, absent any evidence to the contrary, the definition is one of synchronic emergence and therefore does not cover cases in which transformational emergence occurs.

To see how McLaughlin's definition works, consider an example where P is the property of rigidity and the base properties are a specific configuration of quantum states of iron atoms.[12] Assuming for purposes of illustration the fiction that atoms are parts of a macroscopic piece of iron, we can see how the definition might be brought to bear in that case. A duplicate set of atoms in the same basic states would also exhibit rigidity as a matter of nomological necessity. Because it is logically possible that such an arrangement of atoms could lack rigidity in a world in which the laws of physics were different, the property of rigidity does not supervene with logical necessity on the properties of the atoms. However, in our world, the dependency between the micro properties of the parts and the rigidity of the whole is not usually taken to be a fundamental law and so this would not count as an example of supervenience emergence. This requirement of the definition fails to be satisfied because the long-range correlations between iron atoms that give rise to the rigidity of the metal are a consequence at the infinite limit of fundamental principles of quantum mechanics. Because emergence is, on McLaughlin's account, most likely confined to mental properties at best, this result is what one should expect.

Three of our four criteria for emergence will be satisfied by examples that fit McLaughlin's definition. The emergent property P will be novel in the sense of not appearing in the set of properties at the subvenient level, and P will result from the subvenient properties as a consequence of the fundamental law bridging the two domains. Finally,

12. Strictly speaking, the supervening property would be the probability distribution over macroscopic observables, but I shall ignore that difference here.

the supervening property is attached to w as a whole, whereas the base properties are attached to its parts. Hence, the requirement of holism is satisfied—consciousness is usually taken to be a property of the entire neural system, not of its parts. Because supervenience relations are compatible with multiple realizability, the autonomy condition can also be satisfied in some cases, although not in all.

The definition has three central features. The first is that the definition takes the supervenience relation as a relation between properties rather than between predicates. The second feature is the use of nomological necessity rather than logical necessity. The third is the appeal to fundamentality in the definition. Each of these features requires detailed discussion. The first two will be discussed in the next section and the third in section 6.3.

6.2.2 Why Use Supervenience?

Supervenience relations were originally introduced to represent relations between non-natural moral properties and natural properties. The general idea was that if you take two physically and behaviorally indistinguishable individuals in duplicate circumstances (perhaps certain animals such as elephants qualify as moral agents as well as humans), then their moral qualities supervene on the physical and behavioral qualities in this sense: any moral qualities attributed to one individual must necessarily be attributed to the other, physically and behaviorally indistinguishable, individual. This application to moral properties, and later applications to aesthetic and mental properties, had the unfortunate effect of fostering the idea that the higher level properties existed in some way that was distinct from the physical properties that formed the supervenience base, despite the dependence of the former on the latter.[13] The reason was that moral properties, unlike physical properties, cannot be attributed to purely physical objects. A cheese sandwich, desirable

13. As an example of this separation, we have: "*Nonreductive physicalism* . . . consists of two characteristic theses . . . [one of which is] the claim that psychological properties are irreducibly distinct from the underlying physical and biological properties" (Kim 1993b, 339).

though it may be, has no moral qualities even if one can identify morally relevant aspects of its production and place in society.

A second example may be helpful. If B is the property of being a 1966 Volkswagen Beetle and A is the property of being a complete set of parts for that type of car with those parts assembled in the appropriate spatial configuration, then B supervenes upon A. Take another instance of A and, necessarily, you will have a 1966 Volkswagen Beetle. Put another way, when B supervenes on A, B cannot vary independently of A in the following sense: no change can occur in B without some change in A. Put yet another way, necessarily once A is fixed, B is, too.

Beetles do not synchronically emerge from their parts—rather, they are structured aggregates of those parts—and so there is an important difference between this example and the kinds of cases that fall under McLaughlin's definition. In the Volkswagen case, the necessity involved is conceptual, not nomological. One reason for this is that the necessitation follows from the relation between our conception of what it is for something to be a 1966 Volkswagen Beetle and the concepts involved in the basis set A. In fact, one of the virtues of supervenience approaches is that they force us to be clear on what is central to the attribution of the supervening property. In the Volkswagen example, I am assuming that "1966" refers to the model type and year of manufacture of the components, not to when it was actually assembled. The parts must have been made at the appropriate time, because replica 1966 Beetles constructed entirely from parts made in 2015 are not genuine 1966 Beetles. This conclusion is grounded in pragmatic criteria for automobile type identity, similar to the provenance condition required by many for the authenticity of paintings, and there is room for legitimate disagreement about these criteria. Those disagreements do not affect the definition of supervenience—only what must be included in the set of properties A. We do not insist on a similar historical condition for atoms. Any instance of a lithium atom counts as a lithium atom, regardless of when it came into existence.

Here is a test case for the reader. Is it correct to say that we have a 1966 Beetle when the parts are scattered on a garage floor, even though the scattered parts are not in the correct spatial relations to count as forming a car? There is nothing in the world itself that will decide what counts as a correct answer to that question. It is a matter of what we decide counts as a 1966 Beetle, and that decision is about a pragmatic choice of concepts, not facts. If I answer an Internet advertisement for a house for sale and arrive to find a pile of bricks, mortar, window frames, and so on, I may well complain about false advertising, whereas if I had ordered a house from the Sears, Roebuck catalog in 1911, I should have had a different reaction when it arrived.

These conceptual decisions, and the ability to change them, are possible because the Beetle's properties really are nothing but the properties of its parts, suitably arranged. There is no novel ontological entity "over and above" the structured assemblage of parts. This conceptual interpretation falls on the first side of a deep divide in uses of supervenience relations. On the other side lie examples for which a case can be made that the supervening property is genuinely new, distinct from the collection of properties in the subvenient base, and perhaps emergent. The view taken here is that an appeal to supervenience is useful when we are in the first type of situation, when the relation between the supervening and base entities is conceptual, resulting from a difference in representations and there is no ontologically distinct entity at the supervening level. This use of "A supervenes on B" as (ontologically) "A is nothing but B" can be found (in so many words) in Armstrong (1989, 56): "if it supervenes, I suggest, it is not distinct from what it supervenes upon," and in Lewis's:

> A dot-matrix picture has global properties—it is symmetrical, it is cluttered, and whatnot—and yet all there is to the picture is dots and non-dots at each point of the matrix. The global properties are nothing but patterns in the dots. They supervene: no two pictures could differ in their global properties without differing, somewhere, in whether there is or isn't a dot. . . . [S]ymmetry is nothing but a pattern in the arrangement of dots. (Lewis 1986b, 14–15)

The example of moral properties is a possible example of the alternative, ontological view, especially if moral properties are non-natural properties, and there is an argument, not to be ignored, for taking the property approach rather than the conceptual approach in certain cases.

Suppose you are playing a game of Battleships with a whimsical opponent.[14] Unbeknownst to you, he has arranged his ships in a pattern that itself looks like a boat. The pattern of squares that represent his ships on the grid either can be described as having the shape of a boat (call this property B) or can be described by specifying the coordinates of each of the squares (call that complex property A). The reason why, in this case, B is more than a redescription of A is this: a key feature of supervenience is its ability to allow the identity of property instances while maintaining the nonidentity of the properties themselves. The particular instance of A, this particular arrangement of squares, is being described in two different ways: as "boat shaped" and as "square C3 and square C4 and . . . square H8." However, because the property B of being boat shaped can be instanced in a number of different ways, B therefore supervenes on a disjunction of different arrangements of squares.[15] Hence, the properties B and A are different, and there are two properties rather than a single property under different descriptions. This raises the question of in which cases the property approach is the preferred method.

In the uses where reconceptualization does lie at the heart of the issue, supervenience claims can be motivated without a commitment to the separate reality of the supervening properties by using supervenience as a consistency condition. If a concept B supervenes on concepts A, then if two individuals both have A attributed to them, it is inconsistent to attribute B to one but to deny B to the other. This consistency interpretation was the one advocated by Hare (1952, 145), one of the first explicit users of supervenience relations.[16] This use is innocuous,

14. Battleship is a two-player game. Each player has a 10 × 10 grid of squares on which he places battleships of various sizes. Neither player knows the placement of his opponent's ships, and each takes turns calling out the location of a single square on the other's grid. A ship is sunk when all the squares it occupies have been "hit" and the game ends when one player's ships have all been sunk.

15. I have used "instances" rather than "realizations" here because there is no difference in the material type of realization. If the reader prefers the realization terminology, the switch is easy to make.

16. Although Hare is often cited as the originator of the modern philosophical use of the term "supervenient," Lloyd Morgan (1923, 15–16) used the term "supervenient" in what seems to be a way that is similar to contemporary usage, although with downward causation involved. Pepper (1926, 241) also uses the term.

and I shall take it as the appropriate way to construe supervenience. If adopted, it leaves no room for ontological emergence within the supervenience approach. Some users of supervenience wanted something stronger than consistency conditions and shifted the emphasis to a necessitation relation between the two levels of properties. That is when trouble arose.

6.2.3 Supervenience Definitions

Here is one widely used definition of supervenience:[17] "a set of properties M strongly supervenes on a family N of properties if and only if, necessarily, for each x and each property F in M, if F(x) then there is a property G in N such that G(x) and necessarily if any y has G it has F" (see Kim 1993a, 65). We see here the reference to properties in the definition of strong supervenience. Let us call this the property approach and the alternative in which the supervenience relation holds between concepts we shall call the conceptual approach.

The property approach introduces a presumption that the properties F and G are real and that G is not identical to F. For when F and G are the same property, the supervenience relation will collapse into the identity relation, and that would violate the dependency relation that many advocates of supervenience require as part of the relation. An identity relation is not a degenerate case of a supervenience relation. Here are some reasons why. It is appropriate to consider a circle as a degenerate case of an ellipse, with eccentricity zero. To do so facilitates a neat classification and allows an orderly taxonomy within the theory of conic sections. In contrast, a definition of the divisor operation that does not exclude division by zero is defective, not the least reason being that there is a discontinuous change in properties of the division operation when the second argument is zero. A similar type of discontinuity holds in the case of the supervenience relation. A supervenience relation is a determination relation. In contrast, when A and B are identical, A does not

17. Two other commonly used definitions are of "weak supervenience" and "global supervenience." We do not need those definitions here, but interested readers can consult Kim (1993a) for details.

determine B, it *is* B. A supervenience relation is intended to be an asymmetric relation; the identity relation is not.[18]

There is no mention in standard definitions of supervenience of the temporal relationship between the higher and lower levels. With properties, this omission is appropriate, since they are (usually) universals and hence considered as atemporal, but in applications to mereological supervenience, the lack of temporal specification raises the question of whether supervenience relations relate simultaneously to existing objects or whether the supervening object can exist earlier or later than the subvenient object. The standard assumption is one of synchronicity, but global versions of supervenience allow the entire history of a world to form the supervenience base, thus building in access to past states. It is easy to extend these definitions to take into account processes. Here is one possibility:

> *Definition* An individual x of type X *dynamically supervenes* upon a class of objects $Y = \{y_1, \ldots, y_m\}$ just in case necessarily for any object x of type X, for any time t, there are objects $y_1(t), \ldots, y_n(t)$ belonging to Y of types $Y_1, \ldots, Y_n$ and some relations $R_1, \ldots, R_s$ and properties $P_1, \ldots, P_n$, such that $R_1(y_1(t), \ldots, y_n(t)), \ldots, R_s(y_1(t), \ldots, y_n(t))$ and $P_1(f_i(y_1(t), \ldots, y_n(t)), \ldots, P_n(f_j(y_1(t), \ldots, y_n(t)))$ hold for suitable projection functions f_i, f_j, . . . and necessarily, whenever any such $y_1(t), \ldots, y_n(t)$ exist and $R_1(y_1(t), \ldots, y_n(t)), \ldots, R_s(y_1(t), \ldots, y_n(t))$ and $P_1(f_i(y_1(t), \ldots, y_n(t)), \ldots, P_n(f_j(y_1(t), \ldots, y_n(t)))$ hold, x exists.

Next, even when the supervenience relation holds between properties, the definition is formulated in terms of property instances or realizations and the definition requires that the same object *x* has both the subvenient and the supervenient properties. That is to say, if F and G occur at different levels in the hierarchy of properties, and all the higher level properties arise from their supervenience relations to lower level properties, then these definitions of supervenience do not address the possibility that it is new emergent objects that have the higher level properties.

18. See, among many examples, Honderich (1995, 860).

This aspect of synchronic generative atomism can be seen in the restriction in part (1) of McLaughlin's definitions regarding "properties the parts of w have taken separately or in other combinations." Because this is part of a necessary condition for P to be emergent, cases of transformational emergence will not satisfy the definition and will be excluded, even though they are examples of ontological emergence. Transformational emergence is compatible with the emergent property supervening on the parts of w when they are embedded in w as a whole, but those parts when outside the whole, or in other combinations, may not determine the properties of the system as a whole.

6.2.4 Nomological or Logical Necessitation?

The second choice point in the definition of supervenience concerns the strength of necessitation involved. If we restrict ourselves to the same interpretation for both occurrences of the necessity operator in the definition of strong supervenience, we see that when the relation is between properties, the modal operators used in the definition can be interpreted as (a) logical necessity, (b) metaphysical necessity, (c) nomological necessity, and (d) conceptual necessity. The differences between these different strengths of necessitation are a subject of considerable controversy, but fortunately we can sidestep many of the disputed areas. Although McLaughlin's definition requires that emergent properties supervene with nomological, but not with logical necessity on the parts of the whole, we do not lose much by interpreting "logical necessity" to mean either logical necessity, narrowly construed, or metaphysical necessity or conceptual necessity.

> *Definition* A *logically necessitates* B, narrowly construed, just in case the truth of some set of sentences Φ representing A ensures the truth of a set of sentences Ψ representing B as a result solely of the meanings of the logical vocabulary contained in Φ and Ψ.

And,

> *Definition* A set of properties Q *metaphysically necessitates* a property P just in case in all possible worlds in which Q is instantiated, P is also.

Metaphysical necessity is distinguished from logical necessity in the narrow sense because some cases of metaphysical necessity, such as "Nothing can be simultaneously red and green all over its surface," are not truths of logic, and so are not narrowly logically necessary.

> *Definition* A set of properties Q *conceptually necessitates* a property P just in case the intensions of our representation of Q and of our representation of P make it inconsistent to assert that Q can be present without P.

For example, the property of being a whale conceptually necessitates the property of being a mammal, since the intensions of the terms "whale" and "mammal" in standard English make it inconsistent to assert that there can be a whale that is not a mammal.

And, finally,

> *Definition* A set of properties Q *nomologically necessitates* a property P just in case in all possible worlds with the same laws of nature as our own, whenever Q is instantiated, so is P. This is much weaker than metaphysical necessitation because if the laws of nature change, it is metaphysically possible to have Q without P.

In the case of emergence, as an alternative to McLaughlin's definition, we could argue that conceptual emergence can be captured in terms of supervenience by restricting the necessitation involved to conceptual necessity. One way to understand a conceptual approach to emergence is, thus, to consider the predicate describing the emergent property as a reconceptualization of the base properties. This has considerable plausibility if we take emergence to be the result of our inability to construct descriptions of higher level properties ab initio from descriptions of ontologically more basic properties. Under that view, what we are doing when we consider the property of a flock as emergent is reconceptualizing the group of birds not as a collection of individuals but as a whole. This reconceptualization can result in a much more efficient set of predictions of the spatial trajectory of a flock of birds than would be possible if we attempted to make the same predictions starting with the movements of the individual birds. Ontologically, under this view, there is nothing to the flock other than the individual birds and their spatial relations; there is merely a redescription using the language of flocks.

However, the reconceptualization approach to supervenient properties is not possible when the necessitation relation is nomological and an ontological interpretation of supervenience based on a nomological relation involves a much more substantial realist commitment than does the consistency interpretation. One reason it is crucial that these two interpretations of supervenience be kept separate is that the problem of downward causation arises within supervenience accounts of emergence only under the realist interpretation within which the supervening property is not identical to the base properties.

6.2.5 Supervenience Is Not Explanatory

It is one of the unfortunate consequences of earlier work on emergence that in order for a phenomenon to be emergent it must be unexplainable. This is evidenced not only in the examples of emergence that were used at the time, such as life and chemical phenomena, but also in Broad's view that emergent transordinal laws were brute, unexplainable features of the universe, a view that is maintained in McLaughlin's account. It is also a common objection to supervenience relations that they present the relation between the supervening and subvening properties as an unexplained fact about the system.[19] As a result, philosophical theories of emergence based on supervenience relations do not provide an understanding of why that relation holds. This is perhaps why McLaughlin's and other supervenience-based definitions of emergence have only mysterious, currently unexplainable elements of reality, such as qualia and other conscious features of mental life, as emergent. In contrast, the striking thing about most of the scientific

19. For arguments against supervenience relations as explanatory, see Horgan (1993).

examples of emergence described in section 1.8 is that they *can* be explained. This suggests that relying on inexplicable supervenience relations to understand emergence is a result of treating the relations between the emergent phenomenon and what it emerges from at too abstract a level. In scientific explanations, details are usually important, and it is not surprising that when they are omitted, explanations evaporate. Even with mathematical explanations, mere dependency is insufficient for understanding.

6.2.6 Humean Supervenience

The doctrine of Humean supervenience comes in two forms. The first is asserted to be necessarily true and is open to straightforward counterexamples. I am concerned here only with David Lewis's contingent version of the doctrine, since if that is false then those other accounts also fail. Here is Lewis's version:

> [Humean supervenience] is the doctrine that all there is to the world is a vast mosaic of local matters of particular fact, just one little thing and then another. (But it is not part of the thesis that these local matters are mental.) We have geometry: a system of external relations of spatiotemporal distance between points. Maybe points of spacetime itself, maybe point-sized bits of matter or ether or fields, maybe both. And at those points we have local qualities: perfectly natural intrinsic properties which need nothing bigger than a point at which to be instantiated. For short: we have an arrangement of qualities, and that is all. There is no difference without difference in the arrangement of qualities. All else supervenes on that. (Lewis 1986a, ix–x)

The account given in the passage just quoted is not an example of generative atomism, although the two have intellectual sympathies. Humean supervenience denies that emergent properties exist, so even if emergent properties occur several levels up the hierarchy of properties, then Humean supervenience is incorrect.

As an example of a system within which Humean supervenience is plausible, consider cellular automata, of which an example was given in section 4.2. To recap, cellular automata are dynamical systems consisting of an N-dimensional array of cells capable of being in M different states. For clarity, we can consider the case where N = M = 2. The state of each cell is an intrinsic property of that cell, the cells and their states are, within the discrete space and time of cellular automata, point-like, and the rules that update the state of a cell are all local, depending only upon the state of that cell and the states of its nearest neighbors.[20] This last feature is especially important in the context of emergence, because standard cellular automata are inherently nonholistic—all of the computational updating is done on contiguous cells. Cellular automata have become a contentious issue because of claims that they can serve as the basis for a complete computational physics, but we do not need to address those issues here.[21] Thus, if our universe were itself a cellular automaton, then our world would be a Humean world. But that subjunctive possibility should be not taken seriously, for there is no evidence that our world is a cellular automaton and Fredkin (1990), which is often cited as a canonical source, merely provides a sequence of programmatic suggestions for how a cellular automaton might reproduce, approximately, certain features of contemporary physics.

Lewis recognized that there might be physical counter-examples to the Humean supervenience position, but he dismissed them with the remark: "But I am not ready to take lessons in ontology from quantum physics as it now is. First I must see how it looks when it is purified of instrumentalist frivolity . . . and when it is purified of double thinking deviant logic and . . . when it is purified of supernatural tales about the

20. This characterization applies to standard cellular automata. There are other varieties with nonlocal rules, evolutionary rules, and other modifications. I have chosen two-dimensional standard cellular automata because they are easiest to understand. Whether nonlocal rules violate the criteria for Humean supervenience is hard to say, but since their functional equivalent, laws, are merely organizational devices for Lewis, if the use of nonlocal states is acceptable, then so is the use of nonlocal rules.

21. Wolfram (2002) makes some controversial claims along these lines. For a critical review, see Aaronson (2003).

power of the observant mind to make things jump" (1986a, xi). I have argued elsewhere (Humphreys 2013) that this position is unjustified, and I add here only that these quantum mechanical features are the source of macroscopic phenomena that are directly observable. For example, the phase transitions that give rise to superconductivity and superfluidity in helium are a direct result of quantum entanglements. Further arguments along these lines can be found in Shimony (1993, 221).

6.3 FUNDAMENTALITY

As we have seen, many contemporary philosophers hold that emergence is rare and that, at most, conscious mental states are emergent. A much-discussed argument by David Chalmers (1996, chap. 2, section 5) in favor of dualism supports this conclusion, and is one of the best articulated positions in this anti-emergentist tradition. I shall show that this argument, while undeniably instructive, is not effective against the kind of diachronic ontological emergence for which we have argued. Our discussion will also tell us what it is for something to be fundamental and why closer attention needs to be paid to diachronic aspects of emergence.

Chalmers argues that all higher level natural facts in our world logically supervene on micro-level physical facts, perhaps along with microphysical laws. (By "fact," Chalmers means the spatiotemporal distribution of property instances.) Here is his central claim:

> At the global level, biological properties supervene logically on physical properties. Even God could not have created a world that was physically identical to ours but biologically distinct. There is simply no logical space for the biological facts to independently vary. When we fix all the physical facts about the world—including the facts about the distribution of every last particle across space and time—we will in effect also fix the macroscopic shape of all the objects in the world, the way they move and function, the way they physically interact. If there is a living kangaroo

> in this world, then any world that is physically identical to this world will contain a physically identical kangaroo, and that kangaroo will automatically be alive. . . . Given all that information [about the location of every particle in the universe] it has all the information it needs to determine which systems are alive, which systems belong to the same species, and so on. As long as it possesses the biological concepts and has a full specification of the microphysical facts, no other information is relevant. (Chalmers 1996, 35)

This position adheres to the non-separatist supervenience tradition that supervening features do not add to our ontology. As Chalmers puts it, "If B-properties are logically supervenient on A-properties, then there is a sense in which once the A-facts are given, the B-facts are a free lunch. . . . They may be *different* facts (a fact about elephants is not a microphysical fact), but they are not *further* facts" (1996, 41).

In addition to biological facts, chemical facts and nonfundamental physical facts are all claimed to globally supervene on fundamental physical facts. The use of a logical necessity in the supervenience claim seems surprising, because the intuitive reaction is to think that the relation between the two levels should be weaker—most likely one of nomological necessitation. Considering this argument in the context of McLaughlin's definition of emergence that was discussed in the section 6.2.1, according to which a property P of w is emergent if and only if P is nomologically but not logically supervenient on properties of w's parts, suppose we take a property instance to be supervenience emergent if the property itself is supervenience emergent. More specifically, take a property instance [P, a, t] to logically supervene on another property instance [Q, a, t] if P logically supervenes on Q. Then, if all nonfundamental properties logically supervene on fundamental properties, no nonfundamental property instances are supervenience emergent. (Note that neither fundamental properties nor fundamental property instances are supervenience emergent). This link to the positions of McLaughlin and van Cleve explains why Chalmers's claim that only consciousness is emergent seems persuasive. In a later assessment of emergence, Chalmers (2006, 247) asserts that in contrast to other features, consciousness nomologically supervenes on the physical domain.[22]

The degree of surprise at Chalmers's conclusion may be lessened once we note that the microphysical laws are included in the supervenience base, with the result that differences between logically possible worlds in the way that micro-level entities interact with one another have already been excluded. Chalmers is also concerned only with the ontology of our world, and is not committed to the claim that the logical supervenience of higher level facts on fundamental facts is a necessary truth. Since examples of ontological emergence do exist in our world, it is therefore important to see which premises of Chalmers's arguments are false. The primary assumptions of his arguments are that (a) the higher level facts globally supervene on the microphysical facts, where the entire spatiotemporal history of microphysical facts in the universe constitutes the base facts; (b) the fundamental laws as well as specific facts are included in the base; and (c) the microphysical facts are the fundamental physical facts.

Chalmers has three lines of argument in favor of the logical supervenience of higher level facts on physical facts. They are the inconceivability argument, the epistemological argument, and the analyzability argument. I shall focus here mostly on the inconceivability argument, with brief remarks about the epistemological argument. The inconceivability argument asserts that it is inconceivable that there could have been a world indistinguishable from our world in its fundamental

22. His own approach to emergence relies on a distinction between strong and weak emergence, in which weak emergence is characterized in terms of the unexpected nature of truths about the phenomena in question. This is too vague a criterion on which to base a taxonomy, and it suffers from the defects of the "surprising nature" approach of Ronald, Sipper, and Capcarrère (1999). In contrast to many authors who take "strong emergence" to involve downward causation, Chalmers characterizes strong emergence as "A high level phenomenon is *strongly emergent* with respect to a low level domain when the high level phenomenon arises (in some sense) from the low level domain, but the truths concerning that phenomenon are not deducible even in principle from truths in the low-level domain" (2006, 244). This definition allows that the kind of undecidable propositions discussed in section 1.6.2 correspond to strongly emergent physical phenomena. Although Chalmers allows that weak emergence exists, he seems to reject stronger forms of emergence (see, e.g., Chalmers 1996, 129–30, 168n41).

physical features but that was different from ours in its biological, chemical, or other higher level features. To see how Chalmers's argument relies on a tacit assumption, consider the Checkers World of section 1.4.

Let us assume that in Checkers World, at some time N* + 1, all the particles are blue or red particles. At that point, no particle is such that its properties synchronically supervene on the "microphysical facts" of the world. They do not synchronically supervene on the properties of any black or white particles, because there are none, and the laws that govern blue and red particles are not determined by the laws governing the black and white particles or by the particular black and white histories up to N*. More strongly, the properties of the Checkers World at N* + 1 and subsequent times do not globally supervene on the microphysical facts about white and black particles. The blue and red particles are new types of particles that are not composed of elementary particles. They behave differently from the black and white particles—they can move backwards. The laws governing the behavior of the blue and red particles do supervene on the complete set of states of the system, including the states of the blue and red particles themselves, but they do not supervene on the states of the original black and white particles and their laws up to time N*, nor do they globally supervene on the history of the states of the original particles alone.[23]

An appeal to global supervenience, in allowing the entire history of the world to be used as the supervenience base, erases the important distinction between coming to know the laws that govern emergent entities only through induction on their behavior after they have emerged, and knowing it by deduction prior to the appearance of those entities. Although I have put the point epistemically, it can also be made ontologically: the regularities that on a Humean view constitute the laws governing the emergent entities appear only subsequent to the regularities exhibited by the microphysical entities. Lumping the two together within a global supervenience formulation, or through an appeal

23. Strictly speaking, the entire history of Checkers World underdetermines the laws for the blue and red particles, because more than one regularity is consistent with that history. I shall assume we can recover the supervenience of the laws by imposing some pragmatic criteria, although this is not a procedure I endorse.

to limit science, erases a crucial distinction upon which diachronic emergence rests.

Next, because the laws governing blue and red particles do not determine moves—just as in checkers, they determine only what is and is not possible—there can be two worlds, each with identical black and white particle histories up to time N*, but with divergent blue and red particle futures.[24] In fact, this situation will be possible when there is only a single red or blue particle, as long as that particle has the choice of two alternative moves at some point. We could even enhance the game to supercheckers so that there can be brown particles, green particles, and so on, each with different laws for the types.[25] As the world evolves, so too novel domains will develop, none of which supervenes solely on facts about the original black and white "fundamental" particles.

Of course, it could be said that the "physics" of this toy world will simply expand to include red, blue, green, and brown particles, and so on. Fair enough, but then the ontology of this world no longer consists in original physicalism alone; it also includes new domains that develop over time, and the facts and laws about these new domains will not be fixed by the original fundamental facts. So, the claim that everything logically supervenes on the fundamentally physical depends on the assumption that the dynamics of the world do not produce new, nonfundamental and noncomposite entities with indeterministic futures at some point in the history of the world. But, in fact, contemporary physics tells us that such entities do exist in our world and so these conclusions about Checkers World can be applied to the world we inhabit.

Now, ask the question: Are facts about the blue and red particles fundamental facts? Using the distinction introduced in section 2.2, X is fundamental$_S$ if and only if it is noncomposite. But if we adopt that meaning of "fundamental," it prevents, by definition, the Checkers World from counting the blue and the red objects and their properties as emergent. When any noncomposite entity is automatically included in the fundamental ontology, regardless of when it first appeared, the case

24. We can suppose that the moves are made by a physical randomizing device.

25. There is a variant of checkers called Tiers, in which this is possible, but its rules are more complicated than the game described here.

is immovably prejudiced against certain dynamic forms of ontological emergence. Yet this kind of dynamical situation is exactly what transformational emergence claims is how one type of ontologically emergent properties, objects, and states occurs.

The red and blue particles of Checkers World are thus fundamental$_S$ but not fundamental$_d$. Then, call laws that apply to fundamental$_S$ entities, fundamental$_S$ laws, and those laws that are in operation at the temporal origin of the system fundamental$_d$ laws. Chalmers's argument does not exclude transformational emergence unless all of the fundamental$_S$ entities are also fundamental$_d$.

We can also see why Chalmers's inconceivability argument succeeds. Within an ontology in which all and only noncomposite entities are fundamental$_S$, any nonfundamental$_S$ and hence composite, entity such as a classical gas or a population of zebras will be decomposable into fundamental$_S$ entities. It is, therefore, not surprising that it is impossible for a physicalist to conceive of two worlds, in each of which the fundamental$_S$ facts are fixed and in which the fundamental$_S$ laws remain the same, but that differ in their higher level facts. Any noncomposite entity that diachronically emerges over the total history of the world will, as we have seen, automatically be classified as fundamental$_S$ and hence not emergent. The global supervenience approach, itself predicated on an atemporal block universe ontology, converts what is an essentially diachronic type of emergence into a globally synchronic non-emergence.

The second of Chalmers's arguments for the logical supervenience thesis, the epistemological argument, rests on the idea that we have no access to biological facts other than through the physical facts; there is no independent epistemic access to the higher levels. That there is something wrong with this argument should be evident from the fact that, within the physical domain, we have direct access to macroscopic physical facts without knowing what the associated fundamental$_S$ physical facts are. It might be said that this argument simply points to a division between the physical and everything else, including the chemical, the biological, and so on, but even granted that point, our epistemic access to the biological is not through the fundamental$_S$ physical facts but, rather, through nonfundamental$_S$ physics and chemistry. So, the correct starting point for an epistemic argument will be the nonfundamental$_S$ physical facts, and not the fundamental$_S$ physical facts.

The epistemological argument cannot, however, be cleanly separated from the conceivability argument. Ask yourself: Can you conceive of two samples of iron, both originally magnetized and then heated so they pass through 770 degrees Celsius in the absence of an external magnetic field, with one retaining its magnetism and the other losing it? Of course, you can, if magnetization is simply the familiar phenomenological property from elementary school days. But this is no counterexample to Chalmers's claim, because he requires that the fundamental$_S$ physical laws be kept fixed in both cases; and if we fix those, it is (at least) nomologically impossible for the situation just described to occur.[26] Considerations such as these require us to treat even familiar, everyday examples as only prima facie candidates for emergence, and the decision about whether to classify them as emergent will have to be made at the theoretical level, which in this case will involve using models of micro states of the system.

As Laughlin (2005) has noted, our primary reason for believing in the existence of certain macro- or meso-level facts is our ability to directly measure them, rather than on the basis of theoretical derivations from fundamental theory; and the independence of the higher level laws allows us to know things without having to know the most fundamental facts about physics. In that epistemological sense, the most fundamental laws are irrelevant. The use of renormalization and effective field theories also provides grounds for asserting the epistemological autonomy of higher level domains in physics (see Schweber 1993). This autonomy of the higher level laws stops the explanatory regress in law-based accounts of explanation—you do not need to appeal to laws at a lower level to explain the laws at a given level.

Chalmers's epistemological argument can also be evaluated using the Checkers World, by noting that we have direct access to the red and blue entities, rather than access to them being available only through the original black and white entities. This shows that the epistemological

26. Since ferromagnetism is a statistical phenomenon, strictly speaking the claim needs to be probabilistically hedged, but this does not affect the point about the biological case.

argument cannot be a general argument for establishing the logical supervenience thesis. The laws governing the new particles can be inductively inferred from observing the behavior of the red and blue particles, but they cannot be inferred by observing the behavior of the black and white pieces alone, a characteristic feature of emergent entities. Knowledge of the laws governing the elementary particles will leave undetermined what the laws are that govern the emergent particles. We can also see that these laws cannot be determined simply by the way the actual world plays out. There are many possible moves allowed by the laws that never become actual, and so the laws are not determined by the sequence of actual moves.

6.4 MULTIPLE REALIZABILITY

In many cases in which E emerges from F, E will be autonomous from F. The kinds of situation that in philosophy fall under the concept of multiple realizability can provide this kind of autonomy. They represent situations in which, at least to some extent, the state of the system is not dependent upon the specifics of the realization of that state. It is often argued that when a property G is multiply realizable, the multiple realizability prevents the reduction of G and is a reason to employ dependency relations between G and its realizers. I shall argue in this section that these reasons are far from conclusive.

First, we must distinguish between multiple realizability and multiple instantiability. First-order properties are instantiated by individuals. I instantiate the property of being human—so do you, and so do billions of others. This is *multiple instantiability,* and it is (almost) a defining feature of universals that many different individuals can possess that universal. The caveat "almost" is here because there are properties, such as "being a finite number which, when multiplied by any other number returns itself," which are necessarily instantiated by unique individuals—in this case, zero. It would be arbitrary to deny such properties the status of universals.

In contrast, certain kinds of properties can be realized by other properties. This can occur when a property is a functional property, as we saw in section 5.4 on functional reduction. There, a property G realizes a functional property F just in case instances of G stand in appropriate cause and effect relations to instances of other properties so that the functional role of F is carried out. Some properties can be realized by more than one property, and in such cases we have *multiple realizability*. Consider the case of doorstops. These are functionally characterized and the property of being a doorstop can be realized by the property of being a brick, by the property of being a cast-iron pig, by the property of being a wooden wedge, by the property of being a bean bag, and so on.

More generally, consider the concept of a second-order property: "F is a second-order property over a set of base properties B if and only if F is the property of having some property P in B such that D(P), where D specifies a condition on members of B. . . . We may call the base properties satisfying condition D the *realizers* of second-order property F."[27] In the functional case, D(P) will specify causal relations in which P stands. We can then say that a second-order property F is multiply realizable just in case more than one first-order property realizes F. Multiple realizability by itself does not entail that the type of properties that realize the second-order property vary greatly. For that to happen, the types of realizers must be sufficiently heterogeneous that there is no unifying property at the lower level.

It will, therefore, help to split the category of multiple realizability into two types. Let us say that we have *heterogeneous realizability* when the type of entity that realizes the second-order property varies widely. When this variety is not present, type–type identity relations of the kind that multiple realizability arguments are intended to preclude would be possible. When there is more than one type of realizer, but they can all be unified under a single nontrivial type at the same level as the more specific realizers, we have *homogeneous realizability*. There would be little or no point in introducing a dependency relation between the second-order property and its realizers if the property was homogeneously realized, because one could use an identity relation instead of the dependency relation. Thus, the argument for the lack of reducibility

27. Kim (1998, 20).

in cases of multiple realizability rests on the further assumption that the members of B produce heterogeneous realizability. This assumption fails in a number of cases, and it is important not to take prima facie heterogeneous realizability as sufficient for ontological nonreducibility.

Take the multiply realizable property of being a doorstop that was discussed earlier. There is much apparent heterogeneity here. Yet, the relevant causal factor has been isolated by physics as a general property—the property of providing a counteracting force of magnitude at least as great as that exerted by the door. Note that the original general functional property of "a device for holding a door open" results in an analytic definition of the term "doorstop" (assuming we agree that this is an appropriate definition), whereas the property given in terms of the counteracting physical force does not. It is not logically necessary that at least equal and opposing forces keep doors open, because the force laws of classical mechanics might have contained asymmetries favoring forces operating in some preferred direction. It is also not logically necessary that there be only a single general type of cause underlying the action of all doorstops. We could have lived in a world without forces and in which doors were kept from closing by placing spherical objects perpendicular to the door or tetrahedral objects parallel to the door, and friction might have produced a steady flow of heat instead of an opposing force.

It might be argued in response that this general unifying property of providing a counteracting force of magnitude at least as great as that exerted by the door is itself a second-order property, rather than a first-order property, because there are many specific forces including mechanical (such as friction, torsion, leverage), electromagnetic, and electrostatic, each of which realizes the second-order property. As a result, nothing has been achieved. The specification of the second-order property thus constitutes an important unifying step, but it does not do so in a way that results in reduction, nor for that matter, in emergence.[28]

28. I also note that the characterization in terms of a counteracting force is different from what is involved in appealing to the completely unspecific Ramseyfied second-order property $\exists P\varphi$ that is often used by functionalists to describe functional properties.

To assess this reply, consider a more detailed case. It was once common to characterize acids functionally, as substances that produced carbon dioxide when added to sodium or calcium carbonate, or that turned red phenolphthalein colorless. With that characterization, the property of acidity is realized by the property of being acetic acid, by the property of being sulfuric acid, by the property of being hydrochloric acid, and so on, but there was no obvious unifying property possessed by all the realizers, aside from the causal effects on sodium carbonate and other chemical substances. The contemporary Lewis concept of an acid, in contrast, defines an acid as a species that can accept an electron pair.[29] In this approach, we have a unifying common chemical property that is realized by (the property of being) acetic acid, by sulfuric acid, and by hydrochloric acid. In virtue of this common feature, we have a unification of a wide variety of functionally characterized substances to a common chemical property. In this case, one might say that a reduction has taken place—a reduction of the chemical property of being an acid to the strictly physical property of electron pair acceptance. If so, there is no autonomy of the original property of acidity from the now identified physical property of electron pair acceptance, no move to a higher level, and certainly no emergence. Functionalizing the property of acidity is also an unnecessary step because it turns out that we can identify the property of acidity with a physical property. Finally, the distinction between first-order and second-order properties is a logical distinction, and that ordering need not reflect the ontological ordering of properties, as can be seen by the fact that the original functional characterization of acids F was in terms of observable properties and the unifying property E of accepting an electron pair involves an unobservable that is realizable by other unobservable properties A_i. Thus, F is a second-order functional property realized by the A_i that are also realizers of E.

In the case of processes rather than properties, the fact that a dynamical process is independent of the details of the underlying system is not by itself a reason to attribute heterogeneous realizability to the global

29. Two other contemporary concepts of an acid are the Bronsted-Lowry and the Arrhenius conceptions.

properties. There must be distinctively different types of individuals whose dynamics produce those global properties. To take a simple example, in flocks of birds the individual birds can change positions in the flock and different individuals of the same species rotate in and out of the flock, so that we have replacement autonomy; but to have heterogeneous realizability, different species must produce the same flock formation, resulting in equivalence class autonomy. It is the fact that both Canada geese and brown pelicans fly in V formations that makes that type of formation heterogeneously realizable, at least potentially.

6.4.1 Token versus Type Identity

The heterogeneous realizability of many properties has led many philosophers to subscribe to a token identity thesis while rejecting a type identity thesis. The token identity thesis says that if a property P in domain *j* supervenes upon a collection of properties Q in domain *i*, then an instance of P is identical with an instance of some logical combination of members of Q. Thus, if a given event is described by both physical and mental properties, such as a region of the brain being in neurophysiological state Q at time t, and an intense state of depression at time t is also present, then that particular mental event or token of intense depression can be identified with the particular physical event or token of the region's being in Q at t. Yet because an occurrence of that same type of intense depression is, in another person, identified with an occurrence of a different neurophysiological state Q′, we cannot identify the property type P with either the property type Q or with the property type Q′. Nor is it usually possible to identify P with the disjunction (Q or Q′) because in many cases the disjunction of the members of Q will not constitute an admissible property in the sense of corresponding to an acceptable predicate of domain i.[30]

30. The unacceptability could result from the disjunction not corresponding to a natural kind, the corresponding predicate not appearing in the most economical unifying theories of the domain, from a metaphysical objection to disjunctive properties, or some other reason.

We can turn this situation into a straightforward dilemma that affects anyone who subscribes to a particular version of the token identity theory. Suppose you subscribe to these theses:

1. Events are property exemplifications [P, x, t], the notation representing possession of the property P by an object x at time t.
2. The event identity criterion is

$$[P, x, t] = [Q, y, t'] \quad \textit{iff}\ P = Q,\ x = y, \quad \textit{and}\ t = t'$$

3. There are events occurring in irreducible domains, such as mental and physical events, that are token identical, but not type identical.

These three theses are mutually inconsistent, for assuming the truth of thesis (1), if [P, x, t] is some mental event and [Q, y, t′] is some physical event, and the two are token identical, then theses (2) and (3) are straightforwardly contradictory, given that denying type identity means denying that P = Q.

Which of the three theses to give up? Thesis (2) is advocated by Kim, in part because it gives identity criteria that are independent of the way that the properties P and Q are described.[31] However, a promising option seems to be to weaken thesis (2) to the condition:

2′. If $P = Q,\ a = b,\ \textit{and}\ t = t',\ \textit{then}\ [P, a, t] = [Q, b, t']$

Here is an argument in favor of the weaker thesis (2′). Consider any properties P and Q standing in the determinate/determinable relation. For concreteness, consider the event of a planet moving in an elliptical orbit at t, [P, x, t]. Because an ellipse is also a conic section, that planet is

31. "According to this account, each individual event has three unique constituents: a substance (the 'constitutive object' of the event), a property it exemplifies (the 'constitutive property' or 'generic event'), and a time" (Kim 1976, 35). For Kim, a "substance" is anything "like tables, chairs, atoms, living creatures, bits of stuff like water and bronze" (33).

moving in a conic section at t, the event [Q, x, t]. Is the event of the planet moving in an ellipse at t the same event as the planet moving in a conic section at t? If you are inclined to agree that the two events are identical qua tokens— perhaps because you hold that the notation on the left side of (2) picks out spatiotemporally located events that can be identified, but that the properties of being an ellipse and of being a conic section are not the same—then you are amenable to rejecting thesis (2) in favor of thesis (2′). The determinate/determinable relation is a special case of the strong supervenience relation, and so unless there is something special about supervenience relations that are not of the determinate/determinable kind, if one is inclined to reject thesis (2) for thesis (2′), one probably ought to reject it in cases where supervenience holds as well.

Perhaps the situation is clearer with examples that are causal in nature, so consider this case. You have two sources of light. One (source A) emits light of wavelength exactly ω_A—call this "gold"; the other (source B) emits light of wavelength exactly ω_B—call this "saffron." There are two detectors (detectors *a* and *b*) that react to any wavelength in the interval $[\omega_1, \omega_2]$—call this interval "yellow"—and each is placed so as to receive light from exactly one of the sources and only light from that source. The detectors are exact duplicates of one another and both ω_A and ω_B belong to $[\omega_1, \omega_2]$. A beam of light from source A hits detector *a* and the detector goes off. At the same time a beam of light from source B hits detector *b* and that detector goes off. Now, we ask "What caused detector *a* to trigger? Was it a token of yellow or was it a token of gold?" A similar question, with "saffron" replacing "gold" exists for detector *b*. Let the property of being gold (having wavelength exactly ω_A) be P, the property of being saffron (having wavelength exactly ω_B) be P′, and the property of being yellow (having a wavelength in the interval $[\omega_1, \omega_2]$) be Q. The token events involved in our questions are then [P, a, t] and [P′, b, t]. We now invoke a principle adapted from Kim's Causal Inheritance Principle:

> If a determinable Q is instantiated on a given occasion by the determinate P, then the causal powers of *this instance of* Q are identical with (perhaps a subset of) the causal powers of P.[32]

32. Compare Kim (1993a, 355).

This principle, although sensible, needs modification because the circumstances in which the determinate causes the effect are relevant. What caused detector *a* to trigger? Argument: it was that actual beam, the beam of wavelength ω_A that hit detector *a*. There was nothing else in the vicinity, it was sufficient, and the detector went off. Even though any wavelength in the interval $[\omega_1, \omega_2]$ would have done it, there was no backup mechanism in the apparatus to provide an alternative wavelength and so if the gold wavelength had not been present, the detector would not have been triggered. So the relevant causal factor in this case was the determinate property, having a wavelength of exactly ω_A.

Now change the situation slightly. The source is one that emits light of different but particular wavelengths that are all in the interval $[\omega_1, \omega_2]$ and the value of the wavelength emitted changes after each emission. The sequence of emissions is completely deterministic. We can then say that if the source had not emitted the beam of wavelength ω, it would have emitted one of $\omega' \in [\omega_1, \omega_2]$, where $\omega \neq \omega'$, and the detector would have been triggered anyway. So the counterfactual "If the light had not been of wavelength ω, the detector would not have been triggered" is false, and so the beam's having wavelength exactly ω is not the cause of the detector going off.

So, the events on the left-hand side of thesis (2), qua causal events, can be considered as causally identical in the second situation but not in the first, and this in turn indicates that qua causal properties, thesis (2) is correct in the second situation but not in the first. Rather than take the first situation as a reason to reject thesis (2), one could refine (2) to include specifics about the type of property involved and, if causal, the conditions under which it holds. I shall not pursue this further, since it would take us too far afield.

6.5 COMPOSITIONALITY AND AGGREGATIVITY

I suggested in section 1.3 that a necessary condition for emergence is the failure of generative atomism. In an important series of articles (Wimsatt 1997, 2000, 2007), Bill Wimsatt has argued that emergence results from the failure of aggregativity, that emergence is compatible with reduction, that emergence comes in degrees, and that many non-aggregative systems are reductively explainable.[33] Wimsatt provides four necessary and jointly sufficient conditions for a system to be aggregative, and these criteria are justified by appeal to examples that scientists identify as emergent. The criteria are:

> Suppose $P(S_i) = F\{[p_1, p_2, \ldots, p_n(s_1)], [p_1, p_2, \ldots, p_n(s_2)], \ldots, [p_1, p_2, \ldots, p_n(s_m)]\}$ is a composition function for system property $P(S_i)$ in terms of parts' properties $p_1, p_2, \ldots, p_n$, of parts $s_1, s_2 \ldots, s_m$. The composition function is an equation, an inter-level synthetic identity, with the lower level specification a realization or instantiation of the system property.
>
> 1. *IS (InterSubstitution):* Invariance under operations rearranging the parts in a system or interchanging any number of parts with a corresponding number of parts from a relevant equivalence class of parts (cf. commutativity of composition functions).
> 2. *QS (Size Scaling):* Qualitative similarity of the system property (identity, or if a quantitative property, differing only in value) under addition or subtraction of parts (cf. recursive generability of a class of composition functions).
> 3. *RA (Decomposition and ReAggregation)*: Invariance of the system property under operations involving decomposition and reaggregation of parts (cf. associativity of composition functions).
> 4. *CI (Linearity)*: There are no Cooperative or Inhibitory interactions among the parts of the system that affect this property. (Wimsatt 2007, 280–81)

Consider the example of building aggregate described in section 1.6.3. It is a striking fact about this material that it only satisfies Wimsatt's criteria when it is distributed in such a way that there are no interactions between the stones, one case being when they are laid out on a flat floor without touching one another. As previously noted, when

33. The second and third points are also argued by Bedau (2008, 447–48).

a large quantity of aggregate is piled into a heap, it has a type of property, pointed out by Wimsatt, that is not possessed by any of the component stones or by many smaller collections—the property of instability. As was discussed in section 1.2.2, a candidate for an emergent property is the occurrence of avalanches in sand piles, a property which emerges from interactions between the grains in the pile, and which can be possessed neither by a single grain of sand nor by a mere aggregate of sand grains, thus satisfying the novelty condition. The property is holistic in the sense of involving groups of interacting grains and the size of these groups, and hence the avalanches, in some cases satisfying a power law. The property is also to a certain extent autonomous from the underlying type of material forming the heap, because piles of rice grains also exhibit similar behavior, although it is worth noting that this feature is also characteristic of aggregates because of the second part of Wimsatt's invariance condition IS (interchangeability of parts from a relevant equivalence class). This feature is what we called equivalence class autonomy in section 4.2.3. The emergent feature results from interactions between the parts of the system, thus violating the condition CI for aggregativity.

We thus have an example of a non-aggregative property in Wimsatt's sense that is subject to a law that is not applicable to the parts of the system. This is not an example of transformational emergence, but it is considered by many to be an example of nomological emergence. There is a useful lesson here: candidates for emergent properties can appear in familiar medium-sized physical objects and not just in exotic systems.

One attraction of aggregation is that when a composite object undergoes change, all the original constituents remain but have been rearranged. This is not so in cases in which interactions between the constituents undermine the aggregative status of the whole. A complex molecule is not just an aggregate of its elementary particles; there are many interactions, such as Coulomb interactions and exchange energies, between the electrons that give rise to the peculiar features of that molecule. A drystone wall is not just an aggregate of stones in the sense of a heap; it is a skillful arrangement of those rocks that allows it to retain its shape, often for centuries, against weather, scrambling hikers, and the weight of countless sheep. It is certainly different from "the sum of its parts" and it has properties not possessed by its constituents. As we saw in section 2.3.4, a heap of sand heated so that it fuses into glass is a different kind of object from the original heap of sand particles and has different properties, because new kinds of bonds have been set up by the energy introduced in the heating, and it is those bonds that make the silica into a new kind of object, emergent from the original aggregate. Similarly, the addition of water to a dry mixture of Portland cement, builder's aggregate, and sand results in a chemical reaction that produces concrete, which has very different properties from its constituents.

Still, many systems that are non-aggregative in Wimsatt's sense will not be emergent under our criteria, because they continue to fall under the methods of generative atomism. Indeed, because aggregativity imposes conditions on the composition operation that are difficult to satisfy, most systems, whether emergent or not, turn out to be non-aggregative. This means that for Wimsatt, most systems exhibit some degree of emergence. His position thus conflicts with the Rarity Heuristic. It is because of this that, despite the undoubted interest of Wimsatt's position, aggregativity is not the right contrast with emergence.

One example of this is that the aggregativity conditions result in some degree of emergence being attributed to any system that lacks structural symmetry. Consider a plumbing system constructed from plastic components of the type readily available in a hardware store. Put together in the conventional order, they will make up a system that has the two properties of filling the tank in your toilet and halting when it is full. As many amateur plumbers have discovered, install a couple of those components in the reverse order, and the system has the first property but not the second, resulting in a system with the property of filling your basement with water. This failure to satisfy the InterSubstitution condition (IS) is common in systems in which functional organization plays a role, and it is one reason non-aggregative properties are so common under Wimsatt's account. The plumbing system just described is one that falls under the conditions for generative atomism, because generative atomism allows non-commutative operations, whereas aggregativity does not. Wimsatt is correct that emergence is not as rare as many have suggested, but it is also not as common as his account permits.

Wimsatt's position draws on a tradition, perhaps more common in biology than in physics, that views functionality as a source of emergent phenomena. For example, Ernst Mayr has claimed that assembling a hammer head and its handle yields an object with emergent properties.[34] A new property, certainly, but one that is theoretically predictable from the properties of its parts, and thus is not inferentially emergent; it is not ontologically emergent using either the criteria of transformational emergence or those of nonstructural properties; nor is the new concept "hammer" needed for the assembled object, since one can, I'm embarrassed to admit, use a spanner to drive home a motorcycle part.[35] Not all functionally characterized systems possess emergent properties. An example of functional aggregativity is a series of flat rocks that form stepping stones across a stream. The aggregate has a function not possessed by its individual constituents, but that function satisfies each of the four conditions for aggregativity, if the class of rearrangements is restricted to permutations of the existing positions of the stones.[36] One can view this case as an example of a quasi-additive function, with the function of each stone being to allow a person to cross to the point of the stream where the stone is located, and the function of the whole is thus just the sum of the functions of its parts.

Although it is possible to view Wimsatt's position as one of ontological emergence, his own strategy is to turn ontological issues into methodological issues, and his account is therefore focused on models, rather than on the phenomena themselves.[37] He uses this approach to make the important point that scientists tend to give a privileged position to

34. Personal communication, August 15, 1996.

35. Wimsatt's approach uses a kind of contextualism. As he says, "This account analyzes emergence in terms of dependence of a system property on the arrangements of the parts and, ultimately on the context-sensitivity of relational parts' properties to intrasystemic conditions" (2007a, 275). It is not, as it is in the present approach, a change in the components themselves. The context-sensitivity involved is a matter of organization and interaction of the components. Wimsatt credits the context-sensitivity approach to Stuart Glennan. For a different kind of contextual approach to emergence, see Bishop and Atmanspacher (2006).

36. Of course, there is a threshold effect encountered when too many stones are removed, but that is an objection to Wimsatt's original amplifier example as well.

37. Wimsatt (2007, 304, 308).

decompositions of systems that satisfy the aggregativity conditions. This is because, for him, emergence and reduction are often relative to idealizations built into our models of the real system and to a choice of decomposition for the system. By choosing the appropriate level of description and preserving the unity of elements below that level, aggregativity can sometimes be achieved: "the appearance of common and unqualified aggregativity is a chimera and is usually a product of uninspected assumed constancies, idealizations, and overlooked possible dimensions of variation" (Wimsatt 2007, 286). This has the effect that we tend to give priority to decompositions that lead to aggregativity, such decompositions being seen as "natural." That is, because some systems are aggregative under one decomposition but not under others, the bias toward reduction can lead people to search among the decompositions until they find one that produces aggregation. The components of aggregative decompositions then tend to be considered as natural kinds. It is this search for a preferred description allowing decomposition that gives Wimsatt's account a component of conceptual emergence.

Wimsatt's observation is correct, but the method discussed is not unreasonable. It is surely appropriate for a reductionist to look for different ways to decompose a system if it turns out that the original choice of decomposition is inappropriate for carrying out the reduction. The fact that there are different ways of decomposing a system does not entail that none of those ways reflects objective components and structural relations in the actual system. This is a point on which scientific realists have repeatedly insisted, and I refer the reader to some of that literature for further illumination.[38]

6.5.1 A Suggested Amendment to the Non-Aggregativity Approach

For Wimsatt, the classification {aggregate, emergent} is a partition. It can be illuminating to use a finer grained taxonomy of properties, at the least {aggregative, reducible but not aggregative, emergent but explainable, inexplicably ("magically") emergent.} It is the third of these that

38. See, e.g., Psillos (1999).

I am pursuing in this work, and we can gain a connection between Wimsatt's non-aggregativity and other types of emergence by adding a necessary condition for emergence to his four conditions. A necessary condition for a property E displayed by a system S to be emergent is that it be novel in this sense: E belongs to a type domain to which none of the properties possessed by the parts of S belongs, up to but not including the time at which E is first displayed by S.

If we impose this condition, of Wimsatt's examples those that remain as candidates for having an emergent property are: the oscillator, a critical mass of fissionable material, and the stability of rock piles. Those that are excluded as candidates for emergence, in addition to those already classified as aggregative, are: volume, amplification ratio in either linear or nonlinear amplifiers, whether real or idealized; computer memory; fitness.

Here's some elaboration. The type domains can be as broad as the properties characteristic of an entire science, such as psychology, or they can be properties characteristic of a subdomain of a science, such as the properties peculiar to condensed-matter physics. The time at which S first possesses E is excluded to leave room for cases of emergence in which a part of S, as well as the whole of S, displays E at that time. (For example, if the phase transition between water and ice produces the emergent property of being frozen, and if, idealizing somewhat, this phase transition was instantaneous, then the parts of the ice possess the property as well as the whole.)

6.6 EMERGENCE AS NONSTRUCTURAL PROPERTIES

Proponents of combinatorial ontologies apply their combinatorialism to properties using the idea of structural properties. Taking that approach as a foil, O'Connor and Wong (2005) provide an account of ontological emergence that is intended to fit our introspective knowledge about consciousness. Their conception of an emergent property is this: "An emergent property is a property of a composite system that is wholly nonstructural. . . . It is the thesis of emergentism that some basic properties are had by composite individuals. . . . Emergent properties are nonstructural properties of composite individuals. We further presume that they arise from and are sustained by underlying micro structures" (663–64, emphases removed). The key idea concerning nonstructural properties is that the constituents possess latent dispositions that result in a manifest and indecomposable property when the constituents causally interact and the complexity of the system reaches a certain level. If one believes that consciousness forms a unity that cannot be further analyzed and results from complex causal interactions in the brain, this view has much to recommend it. However, we have seen with entangled states that quite simple systems can give rise to emergent properties and so the complexity requirement is unnecessary.

I shall show that because the nonstructural approach retains a connection to the generative atomism tradition through its requirement that the emergent property be possessed by a composite system, it is too narrow to include all cases of ontological emergence, and in fact can be considered to be a special case of fusion emergence. To begin, consider what counts as a structural property: "A property, S, is structural if and only if proper parts of particulars having S have properties not identical with S and jointly stand in relation R, and this state of affairs *is* the particular's having S" (O'Connor and Wong 2005, 663).[39]

The definition of a structural property fits the tradition of generative atomism applied to properties, and it reflects the view that structural properties are nothing "over and above" the relational aggregate of the properties possessed by the object's parts. Note that the definition, in making essential use of proper parts of particulars, presupposes compositionality for the object possessing the structural property.[40]

39. O'Connor and Wong's own definition is based on David Armstrong's position on structural universals: "To be an instance of a structural universal F a particular must be made up of just two non-overlapping parts, one of which instantiates universal G while the other instantiates universal H, with the G part and the H part linked by the external relation R" (Armstrong 1978a, 69; see also Armstrong 1997, 153).

40. In contrast, conjunctive properties are not structural. This is because the holding or not holding of conjunctions does not depend upon what relations exist between the individuals involved, even though the existence of certain relations may be entailed by the conjunction. If it is true that my house is large and your house is small, there is a

Taking this definition, the property P of being a suit of diamonds in the standard order within a deck of playing cards is a structural property of the suit of cards, because there are proper parts to a suit of cards—that is, the cards ace through king, having properties such as P_9 "is the nine of diamonds" that are different from P, and when those proper parts stand in the usual ordering relation R, those parts are what it is to be a suit of playing cards in the standard order. Now, contrast this with the property F of being a straight flush in poker (five cards of the same suit with sequential values). Whereas it is essential that the spatial ordering of the cards in a deck following the standard order be respected—otherwise the concept of a shuffled deck would have no content—the spatial order in a straight flush is irrelevant, even if it is conventionally used to display the flush, and so the spatial relations of the cards cannot serve as a basis for F counting as a structural property, whereas the ordinal ranking of the cards' values can provide such an R. This is a sufficiently abstract relation that it would not make F a physical structural property.

So, this requires us to decide whether, in the context of physicalism, the candidates for the structuring relation R must all be physical relations. If we do so decide, then the property of being a straight flush is a nonstructural property, an awkward although not necessarily disabling result.[41] More generally, if T indicates the domain of some science, should we insist in discussions about whether there are emergent, non-T properties that non-T relations cannot structure collections of T objects? That seems reasonable, but then there will be biological objects that have nonstructural properties according to the above definition, but that do have social structure.

One of the virtues of O'Connor and Wong's account of emergence is that it is freed from the need to talk in terms of levels—and indeed, their goal is to provide a causal theory free of the problems induced by downward causation: "If [emergent properties'] appearance in certain systems is to be explained at all, they must be explained in terms of a

physical relation of "is smaller than" that holds between your house and mine, but this does not follow from the semantics of conjunction alone.

41. Because poker, as with most nonathletic games, can be played with tokens of radical different designs and hence is multiply and heterogeneously realizable, providing a physicalistically acceptable solution to this problem would not be easy.

causal, not purely formal, relationship to underlying, immediately *preceding* structures" (2005, 664). This feature places their approach in the category of diachronic emergence while leaving open the possibility of synchronic emergence also being present.

A valuable feature of O'Connor and Wong's analysis is their argument that nonreductive physicalism, a position that they oppose, avoids facing the pressing ontological issues involved in emergence, because nonreductive physicalism allows the token identity of mental and physical events but rejects their type identity. This, as they point out (2005, 660), is problematic because token mental events have properties, such as intentionality, that are distinct from any physical event. Thus, even if the type differences are conceptual rather than ontological, and we have a hybrid account that uses elements of both conceptual and ontological emergence, properties will be attributed to token physical events that many have argued it is impossible for them to possess. This is troubling because for those of us who subscribe to singular accounts of causation, it is the token identity of the causes that is important. O'Connor and Wong are thus correct in saying that making the domain of the mental conceptually irreducible to the domain of the physical while allowing token identity is misleading; from an ontological perspective, all this means is that simplicity considerations lead us to use a new vocabulary once the degree of complexity of a physical system reaches a certain value. (For a related issue, see section 6.4.1.)

In the formulations given by O'Connor and Wong, being structured is primarily a feature of objects and is only derivatively a feature of a property. Indeed, without qualification, O'Connor and Wong's definition allows that the same property is structural in some instances and nonstructural in others. For example, the property "having a neutral charge" will be a structural property when possessed by a composite object such as an atom that has charged parts (the relation R will be sufficient spatial proximity that an overall neutral charge results), and will not be a structural property when applied to a chargeless basic object that has no proper parts and no internal structure. This will not make the neutral charge on the basic object emergent under O'Connor and Wong's definition, but the relativity is a troubling feature of the definition of a structural property. One cannot avoid this problem by banning different values of quantitative properties from counting as different properties, because the qualitative (and physically important) property "having a positive charge" also suffers from this problem. Simply consider, on the one hand, a composite object such as a sodium cation with overall positive charge, but having both positively and negatively charged parts, and on the other hand, a positively charged elementary particle.

6.6.1 The Relation of Nonstructural Properties to Transformational Emergence

Transformational emergence does not always give rise to nonstructural properties. The separate property instances $P(a_1), \ldots, P(a_n)$ that are changed to $P'(a'_1), \ldots, P'(a'_n)$ in virtue of interactions can be compositionally combined to form a new structural property $R(P'(a'_1), \ldots, P'(a'_n))$ that counts as emergent under the view argued for in section 2.2. So, there will be cases of transformational emergence that are not covered by the nonstructural approach. When fusion occurs, in contrast, it gives rise to a nonstructural property in cases where the original objects retain their identity through the fusion operation (even though the properties do not). This is because the fused property cannot be identified with particulars standing in some relation. In cases where the original objects as well as the properties fuse, the definitions of structural and nonstructural properties simply do not apply because there are no proper parts to the system, and so we again have a case not covered by the nonstructural approach. A nonstructural property is thus a particular type of fusion emergent property that is present in cases where (a) there is no ontologically emergent particular, and (b) there is a diachronic nomic sufficiency that leads to the property fusion.

David Lewis (1986c) raised the following argument against structural universals when they are conceived of as mereological sums of constituent universals: Being methane is a structural universal composed of the universals being carbon, being hydrogen, and being bonded. A methane molecule has four hydrogen atoms and so the universal of being methane must contain the universal of being hydrogen four times, but because there is only one universal of being hydrogen, being methane cannot be a mereological sum of other universals.

The choice of methane as an illustrative example is problematic, however.[42] There are two different senses of "structural" that need to be kept separate in discussions of chemical examples. Molecular properties are considered structural by chemists because they depend on the geometry of the molecule. The philosophers' definition requires that there be no overlap between the constituents, but the carbon-hydrogen bonds in methane are covalent and, as we saw in section 2.3.5, the basis for covalent bonding is the overlap between the valence electrons. Indeed, O'Connor and Wong (2005, 663) have to engage in the fiction that carbon and hydrogen atoms are mereological atoms in order to attribute credibility to this example.

Bigelow and Pargetter's (1990) account of structural universals is intended to avoid Lewis's argument and identifies such properties as being methane with a second-order relational property of an individual—in this case "the property of having a part which has the property of being hydrogen and having a part which is distinct from the first part which has the property of being hydrogen and . . ." (87–88). A key feature of their approach is that standing in certain internal relations is an essential characteristic of methane, where Bigelow and Pargetter's concept of an internal relation is closer to an M-internal relation than to an L-internal relation: "The related terms could not exist without standing in those [internal] relations" (88).

With this in mind, we can see that it is important for O'Connor and Wong's definition whether the relation R is an internal or an external relation, and furthermore that we are clear on the sense in which "internal relation" and "external relation" are taken. If we take R as an M-external relation, then we retain the purely structural, combinatorial aspect that is intended by their definition. But if R is an M-internal relation, then the "components" of the individual possessing the structural property are not the same inside the relation as they are outside it, and it is in virtue of those changes that the "structural" property obtains.

42. David Armstrong also appeals to the property of "being a methane molecule" as an example of a structural universal (Armstrong 1997, 34).

6.7 PROPERTIES AND OBJECTS

In the characterization of transformational emergence, I appealed to changes in essential properties of an individual. Readers can choose their own account of essential properties, but it is time to say something about properties in general. The term "property" is multiply ambiguous. Some properties have semantic work to do; others have causal roles to play; some properties are physical and others are abstract. Properties play a metaphysical role as universals, and in picking out natural and unnatural kinds. There is no reason why something that serves one of these functions must thereby serve all the others as well. To take the most obvious difference, semantic properties behave differently from physical properties. The class of semantic properties is usually considered to be closed under the standard Boolean operations of negation, conjunction, and disjunction. There are well-known reasons why the class of physical properties should not be required to be closed under negation and disjunction: there is the property of being an electron, for example, but the predicate "is not an electron" does not have a corresponding physical property; the predicate is satisfied when an entity lacks at least one property necessary for an object to be an electron, and the lack of a property is not, ipso facto, itself a physical property.

It is similar with disjunction: being a rabbit and being over six feet tall are separately physical properties, but being either a rabbit or over six feet tall is not. Some who have denied that physical properties are closed under all Boolean operations (e.g., Armstrong 1978b) have allowed that instantiated properties are closed under conjunction, but this is incorrect if universal closure under logical conjunction is intended, for nomologically exclusive properties cannot be conjoined and instantiated. One such example is the nonproperty of being at 1000 degrees Celsius and being at atmospheric pressure and being liquid water. So nomological coinstantiation must be used instead of conjunction. These considerations all indicate that which algebraic or logical structure is appropriate to represent the ways in which properties combine in the world is a matter for empirical determination, and will differ from type of property to type of property. It is not a matter for a priori decisions.

So, call individuals that possess essential properties and that are structured by causal interactions governed by natural laws "nonaccidental individuals." This restriction immediately avoids many of the objections to bundle theories of individuals.[43] First, some collections of properties fall into the category of accidental individuals, in part because such collections are the result of our imposing conceptual structure on naturally inhomogeneous collections. A garbage dump is one such accidental individual, its collocated contents influenced by whatever the contemporary society decides to discard.[44] Yet the collocations of properties that make up accidental individuals can be multiply instantiated. The number of garbage dumps is large and growing. So, accidental properties can be universals, too.

Second, individuals are contingently existing objects and are not necessarily existent even if the properties that make them up necessarily exist. It is best to avoid speaking in terms of collocated properties being instantiated, because it suggests that a property has to be instantiated by something. This can lead to a felt need for particulars to do the instantiating—a need best done without.

Third, we should not be distracted by the difficulty of identifying the essential properties for a given type of individual. We may have doubts about various candidates for the essential properties associated with humans, but there are no doubts about what it takes to be a hydrogen atom—one proton and one electron suitably distributed in space. Genetics may one day tell us what it is to be an Indian elephant, but we know now that a circus elephant with serious but incomplete memory loss is still an elephant, and hence that a normally functioning memory is not an essential property of an elephant qua biological type.

Fourth, spatiotemporal continuity conditions allow us to distinguish one individual from others that have the same core set of properties. We should note that sometimes a single essential property can distinguish one particular from all others. Every automobile in the United States

43. Many of these objections are clearly laid out in van Cleve 1985.

44. For communicative purposes, I am here and elsewhere referring to complex collections of properties as collections of "individuals" rather than collections of collections of properties. The ontology remains one fundamentally of properties.

carries a unique Vehicle Identification Number (VIN). The property of having that number, by itself, individuates that vehicle from every other. Suppose that, by mistake, two otherwise indistinguishable vehicles are given the same VIN. Then it is only spatiotemporal continuity that will differentiate them. As a matter of fact, in our world they will bear different relations to other property clusters, and so we can thus differentiate them. But this is not what makes them different—the individuating properties must be intrinsic or else the identity of the individual would change in virtue of removing all the nonintrinsic properties it bears to other individuals.

Finally, suppose we describe a property as emergent. Then, it might appear that the predicate "emergent" picks out a second-order property. But "emergent property" has the same form as "fundamental property" or "structural property"; it is a classifier that also applies to objects and remains as a first-order attribute. Emergent properties are thus a subclass of first-order properties, and therefore are different, logically, from realizers, which are second-order properties.

7

Scientific Topics Related to Emergence

7.1 AN EXAMPLE: FERROMAGNETISM

In this section I shall discuss in some detail a possible case of ontological emergence that is different from transformational emergence and fusion emergence. I noted in section 2.1 that many claims about ontological emergence are claims about models of the emergent phenomenon, and that one should be cautious in moving from those model-based claims to conclusions about the existence of emergence in the corresponding natural system. In this section I shall examine in some detail the example of ferromagnetism, which is one of the most cited examples of ontological emergence, and I will try to identify what must be done to establish a claim of ontological emergence. Throughout, the focus will be on the philosophical issues, but it is crucial to appreciate certain details of the scientific theories and models because they allow us to understand how a phenomenon can both be emergent and explainable in a transparent way. The physical interactions that give rise to ferromagnetism will be emphasized, in contrast to other treatments of this and related phenomena that highlight the mathematical aspects. For example, a much more abstract treatment of critical phenomena can be given using the methods of renormalization group theory. I shall not attempt a discussion of that approach here, instead referring the reader to Batterman (2000), Castellani (2002), and related literature.

Many putative examples of ontological emergence occur within the domain of macroscopic systems. A macroscopic system can be defined as one whose equation of state is independent of its size. An ideal macroscopic system is one containing an infinite number of particles, the density of which is finite. A real macroscopic system is one that is sufficiently large that its equation of state is empirically indistinguishable from an ideal macroscopic system.

These definitions are purely objective, and do not rely on a characterization of "macroscopic" by way of reference to human observers. Two of the principal features of macroscopic systems are that they display qualitatively different properties from the micro components that compose them and that they are holistic. Here is a representative view:

> Macroscopic systems enjoy properties that are qualitatively different from those of atoms and molecules, despite the fact that they are composed of the same basic constituents, namely nuclei and electrons. For example, they exhibit phenomena such as phase transitions, dissipative processes, and even biological growth that do not occur in the atomic world. Evidently, such phenomena must be, in some sense, *collective,* in that they involve the cooperation of enormous numbers of particles: for otherwise the properties of macroscopic systems would essentially reduce to those of independent atoms and molecules. (Sewell 1986, 3)

Macroscopic systems thus have properties that are holistic and irreducible, and the emergent properties cannot be possessed by individuals at the lower level because they occur only in the case of infinite collections of constituents. Within the area of macroscopic systems, emergent phenomena are often linked with features such as universality, phase transitions, broken symmetry, self-organization, and scale-free phenomena.

7.1.1 Basic Features of Ferromagnetism

Ferromagnetism appears as the temperature of certain materials such as iron and nickel drops below their Curie or critical temperature T_c. Above the Curie temperature, which for iron is 770 degrees Celsius, the material is paramagnetic—that is, the magnetization of the iron disappears when the external field is reduced to zero. Below the Curie temperature, ferromagnets remain permanently magnetized when an external magnetic field is removed. The presence of a phase transition and the appearance of a novel property—here, the ability to remain magnetized in the absence of an external field—are frequently taken as reasons to count ferromagnetism as an example of emergence.

Consider a ferromagnet, the temperature of which can vary. The temperature is a *control parameter,* a quantity that links the system with its environment, and the net magnetization, which is the thermodynamical average of the spins, is an *order parameter.* More generally, an order parameter is a variable that determines when the onset of self-organization occurs, and it serves as a quantitative measure of the loss of symmetry.[1] In a ferromagnet, the order parameter is the net magnetization M.[2] There is no general method for finding order parameters, and identifying an order parameter is very system specific.[3] The critical temperature, and more generally a *critical point,* is a value of the control parameter at or near which a phase transition occurs and large-scale structure appears as the result of local interactions. At the critical point, scale invariance of the structures exists.

A *phase* is a region of the phase space within which the properties are analytic as functions of external variables—that is, they have convergent Taylor expansions. This means that with small perturbations the system stays within the phase and all thermodynamical properties will be functions of the free energy and its derivatives. A point at which the free energy is non-analytic is a *phase transition.* Looked at from the perspective of perturbations, the non-analyticity means that a small perturbation in the control parameter can lead to a switch from one phase to another and to corresponding large qualitative changes. Indeed, it is with this reference to non-analyticity and the associated singularities that a distinctively different style of explanation for phase transitions is said to enter, a fact that has been emphasized by Robert Batterman (2000, 2002).

1. More technically, an order parameter is a monotonic function of the first derivative of a free energy (van Dyke Parunak, Brueckner, and Savit 2004, 1).
2. Others examples of order parameters are the director D of a nematic liquid crystal, the amplitude ρ_G of the density wave in a crystal, and the mean pair field in a superconductor $<\psi, r>$.
3. See, e.g., Binney et al. (1992, 11).

There is a standard distinction made between two types of phase transitions. In the older terminology, these are called first-order and second-order phase transitions, whereas more recent literature prefers the terms "discontinuous" and "continuous."[4] In first-order transitions, the order parameter changes discontinuously, whereas in second-order phase transitions it is the rate of change of the order parameter with respect to changes in external variables that is discontinuous. The paramagnetism to ferromagnetism transition is a second-order phase transition, and if we slowed down the process, we should see that there is a gradual shift toward organization and order. In contrast, the water-to-steam transition is a first-order transition. Although there are important differences between the two types, for the purposes of ontological emergence, it is of no great importance whether the transition is continuous or discontinuous as long as the beginning and end of the process involve qualitative different properties. In contrast, as Batterman has emphasized, it is the presence of singularities at the critical point that supports the case for theoretical models of some critical phenomena falling into the category of inferential emergence.

In a ferromagnet, atoms having partially filled electron shells are arrayed in a lattice structure. Each of these lattice sites has a magnetic dipole moment—a spin plus an orbital angular momentum. Neighboring dipoles, which act as tiny magnets, tend to align themselves because parallel dipoles produce a lower energy state than do dipoles oriented in opposite directions, this difference in energy being known as the exchange energy. Formally, this state is represented by the pair of dipoles having an anti-symmetric state function, a consequence of Pauli's Exclusion Principle. This parallel alignment occurs spontaneously and is a specifically quantum mechanical effect because, under classical electromagnetism, the lower energy state would result from

4. Sethna (2006) recommends against the terms "first-order phase transition" and "second-order phase transition": "This is not only because their origin is obscure, but also because in the latter case it is misleading: the thermodynamic quantities at a continuous phase transition have power law singularities or essential singularities, not plain discontinuities in the second derivative" (243n7). Some phenomena exhibit continuous phase transitions under some conditions but discontinuous phase transitions under other conditions.

opposite, not aligned, dipoles. Random fluctuations due to thermal variations will initiate a process that determines which orientation the final state will have. I should note, however, that much of the contemporary theory of ferromagnetism is semi-classical, rather than fully quantum-mechanical, in that it treats spin as a classical vector, for example. It is model-dependent fictions such as this that should make one cautious about inferring the existence of ontological emergence on the basis of models.[5]

Order in a system such as this is represented by structure that results from a loss of symmetry in the system.[6] A preliminary word of caution is necessary regarding the use of the term "symmetry." The term is often used in these contexts to refer to a certain kind of homogeneity that is typically produced by a disordered spatial arrangement. Thus, the symmetry in many systems is simply the presence of a random spatial arrangement of the values of a given property. In the ferromagnetic state, iron appears to have a great deal of symmetry, since all the spins are aligned, but that is a completely different and informal use of the term. This ambiguity is confusing, but the uses of the term have become entrenched, and it is essential to be aware of their differences. A key aspect of spontaneous symmetry breaking is that the aggregate order allows the energy in the system to be minimized.[7]

The alignment of spins in ferromagnetism is an example of spontaneous symmetry breaking because the original rotational symmetry possessed by randomly oriented spins is replaced by a directionally preferred state in which that symmetry is no longer present. At high temperatures, the thermal energy is sufficient to overcome the exchange-energy bonds between the dipoles, but as the temperature is lowered, this randomizing influence decreases until at the critical temperature it disappears.[8]

5. See Aharoni (2000).
6. I shall not discuss here topological order, which goes beyond the older Landau symmetry approach that is the basis of my discussion.
7. Some phase transitions, such as the gas–liquid transition, do not involve broken symmetry. See Auyung (1998, 184).
8. Anderson (2005, 423) asserts that the ferromagnetic case is an atypical example of broken symmetry. As an example of emergent phenomena, however, it is widely cited.

The parallel orientations of the dipoles initially occur only in small domains of about 10^6 to 10^9 atoms (known as Weiss domains, the existence of which has been experimentally established) that are dominated by short-range interactions.[9] In order to arrive at the macroscopically observable magnetization of the material, the Weiss domains, each of whose net magnetization is oriented in a random direction, must then be aligned by an external magnetic field. A key distinction for us is between the spontaneous formation of long-range orderings within the Weiss domains, which occurs in ferromagnets in the absence of an external magnetic field, and the alignment or growth of the Weiss domains as a result of the application of an external field.

Thus, M, the net magnetization, is a holistic property of the entire system; it emerges through the formation of the Weiss domains and then spreads throughout the system. Above the critical temperature, the property of a magnetic orientation is present in each of the micro constituents, but a zero level of M is present at the global scale. Below the critical temperature, M becomes non-zero at the global scale.

7.1.2 The Status of Ferromagnetism as an Emergent Phenomenon

In the context of our classification of types of emergence, the inability to predict past a critical point puts phase transitions into the category of inferential emergence, with the dynamics of the process as the control parameter is varied thus placing it in the diachronic half of the division. Yet, the inferential emergence is a result of ontological changes in the system. Up to the critical point, the symmetry breaking is a result of a dynamic version of generative atomism (at least within the Ising model account as described here), with fixed rules governing the spin-flip dynamics. The holistic aspect of ferromagnetism, of which the long-range correlations are a central feature, is a result of local interactions. In these respects, there is nothing mysterious about the appearance of the ferromagnetic state, and there is nothing incompatible with the

9. For example, Silva, Schultz, and Weller (1994).

generative atomism view at temperature values away from the critical point. We must look to the phase transition for that.

An important concept here is the *correlation length*—the maximum distance over which a significantly non-zero correlation between the spins obtains. In the initial, randomly oriented state of the material, the correlation length is essentially zero. As the Weiss domains form, the correlation length gradually increases until, in the final ferromagnetic state, it equals the length of the magnet and is responsible for the holistic features of ferromagnetism. The symmetry breaking and the long-range correlations of the spins produced by it are essentially nonlocal properties.

Taking our four criteria for emergence (novelty, autonomy, one entity resulting from others, and holism) in turn, ferromagnets have a number of novel properties. The most obvious novelty is the property of ferromagnetism itself, which at the micro level reflects the transition from a disordered micro state to an ordered micro state, the result of spontaneous symmetry breaking. This property is autonomous both in the sense that it persists after the external magnetic field has been removed and in the sense that it is largely independent of the details of the microdynamics. What this independence from the microdynamics means is something that requires careful consideration. The usual sense is that ferromagnetism is exhibited by many different materials, and is thus multiply realizable. Multiple instantiability of the overall magnetization is always a feature, because many different micro states can give rise to the same overall magnetization. However, since only the presence of magnetic moments at the lattice sites can lead to overall magnetization, once this is made clear through the abstraction involved in the Ising model, we no longer have multiple realizability.

Multiple realizability by itself does not guarantee irreducibility, as we saw in section 6.4, because physics can sometimes identify a common feature had by all realizations, and this common feature explains the existence of the macro property. It is true, of course, that different materials can be ferromagnets, such as nickel and cobalt in addition to iron, but what this illustrates is that once we understand why multiple realizability can occur, we often can reduce multiple realizability to single realizability.

Regarding other features relevant to emergence, the long-range correlations that underlie the ferrromagnetic property result from the exchange interactions between the spins on the lattice sites, and the fact that the correlation length coincides with the length of the material shows that there is an essentially holistic aspect to this phenomenon. In turn, this holistic feature prevents the reduction of ferromagnetism to facts about the micro level, although the macro property has emerged from the micro features. A further feature is that the magnetization M is a result of the self-organization of the system.[10]

These four generic characteristics of emergence are present because of other features that are often cited in physics as symptoms of emergence: the self-organization of the Weiss domains that is the process by which M results from the dynamics of the spin-flips, the presence of spontaneous symmetry breaking, and the existence of a (second-order, continuous) phase transition from the paramagnetic state to the distinctively different ferromagnetic state. It is the connections between these specific physical processes and the four generic features of emergence that allows the emergent features to be explained, if one is willing to allow that the model of ferromagnetism described here is an accurate description of real ferromagnets.

7.1.3 Models, Possibilities, and Actuality

I noted earlier that any claims about emergence must be careful to distinguish among phenomena that are actual, phenomena that are nomologically possible, and phenomena that are merely logically possible. We shall now discuss some approaches in which this separation is especially important, because the emergence claims are based on models in which an infinite number of objects are present, asymptotic limits are used, or the claims themselves are mathematical in form.[11]

10. Howard (2007) suggests that our current state of knowledge of the relations between elementary-particle physics and condensed-matter physics is insufficient to justify claims of emergence within the latter. I agree that caution is in order, but I hope to have addressed at least some of the concerns raised by Howard.

11. Only the broadest outline of these features can be given here, and for a proper understanding, it is essential to read the primary literature, some of which was referenced in earlier citations.

Phase transitions within mathematical models are mathematical singularities, and this raises the question of whether such singularities correspond to anything that occurs in real systems. Within a theoretical treatment, the transition between pure water and ice is a discontinuous phase transition. Pure examples of water do not exist naturally on earth; even samples from pristine glaciers contain some impurities. Despite the impurities, it would be legitimate to cite the change from pure water to ice as an example of an emergent phenomenon if it satisfied our philosophical theories of emergence, on the grounds that the actual system is a close approximation of the idealized model. The central problem with phase transitions is that they are different from this kind of case involving approximations, because the theoretical results require infinite limits.

A related issue is whether emergent phenomena only occur in systems with an infinite number of degrees of freedom. In the Landau-Hopf model of turbulence, the turbulent state is represented as the superposition of an infinite number of independently periodic motions, and the turbulent phase is only possible for a system with an infinite number of degrees of freedom.[12] In the present case, the fundamental philosophical problem is that ferromagnets have perfectly sharp phase transitions only in idealized models with infinite degrees of freedom. In real systems, there is not a sharp transition between water and ice, but a triple point, where water, ice, and vapor coexist. More realistic models have to use finite systems and asymptotic expansions of the thermodynamic functions. For technical details and a perspective on different viewpoints, I refer the reader to Batterman (2002), Butterfield (2011b), and Callender (2001).

At the observable level, the phase transition from the paramagnetic phase to the ferromagnetic phase appears to be discontinuous The apparent discontinuity of the transition raises a question about whether emergent phenomena should be identified at the observable level or, rather, identified with unobservable phenomena that are accessible only through our theoretical models. If you were to heat an iron magnet to

12. I owe this example to Sara Franceschelli.

its critical temperature, the converse transition from the ferromagnetic to the paramagnetic would appear to be quite sharp, whereas the phase transition of boiling water into steam seems to be a continuous process that takes considerable time. The property of ferromagnetism and the existence of a phase transition thus appear to be observable, phenomenological features of iron, just as does the phase transition from water to water vapor. Yet, theoretical physics classifies the liquid-to-vapor transition as a discontinuous phase transition and the ferromagnetic-to-paramagnetic transition as a continuous phase transition. If a process that appears, observationally, to be discontinuous but is in reality continuous, the demands on the reductive project to reproduce the former property will be inappropriate. This is why the question arises: Should we look to observable or to theoretical criteria to decide which features emergent properties possess?[13] The traditional examples of emergent phenomena—life, consciousness, differences between water and hydrogen and oxygen —all appealed to distinctive, observable or introspective, features of the phenomena to initially identify them. When Mill asserted that no trace of the properties of hydrogen and oxygen could be found in the properties of water, he was referring to macroscopically observable quantities of the chemicals, not to molecular-level features. It is unlikely that we shall find clear examples of emergence in the directly observable realm, with the possible exception of some mental phenomena.

7.2 LINEARITY, NONLINEARITY, AND COMPLEXITY

A great deal of traditional science has been based on analysis. The classic method of laboratory experimentation, for example, is based on the assumption that it is possible to isolate individual causal influences by controlling other variables, and then to combine the results from different experiments. This decomposition and recomposition method is

13. This question is related to a slightly different question: Are phase transitions statistical mechanical or thermodynamical phenomena? For a discussion of that question, see Callender (2001) and Liu (1999).

mirrored in many mathematical models using various versions of additivity principles. In recent years, much attention has been paid to the relations among nonlinear systems, complexity theory, and emergence; and it is often claimed that nonlinear systems form a subset of the set of complex systems, and that complex systems give rise to emergent phenomena. This section is a brief guide to the origin of those claims.

7.2.1 Linearity

The property of linearity is best captured through its use in mathematical models. The term "linear system" can then be applied to systems that are correctly described by a linear model. Suppose that an input to the model at time t is represented by $x(t)$, and that the temporal development of the model is described by an operator O, so that the output of the model at time t', where t' is later than t, is represented by $x(t') = \mathrm{O}[x(t)]$. Then, if the model is presented with two distinct inputs $x_1(t)$ and $x_2(t)$, the system is *linear* just in case $\mathrm{O}[ax_1(t) + bx_2(t)] = ax_1(t') + bx_2(t')$, for any real numbers a and b. This superposition principle—that if f, g are solutions to an equation E, then so is $af + bg$ for arbitrary constants a, b—is the characteristic feature of linear systems, and it represents a specific kind of additive decompositionality.

Nonlinear systems violate superposition. Consider an atemporal case consisting of the wave function given by $\sin(\theta)$. This describes a nonlinear system because in general, $\sin(a\theta_1 + b\theta_2) \neq a\sin(\theta_1) + b\sin(\theta_2)$. As a result, the system cannot be modeled by decomposing the angle θ into smaller angles θ_1 and θ_2, finding the solutions to $\sin(\theta_1)$ and $\sin(\theta_2)$ and adding the results.

When using this definition to represent a natural system, one must be careful to specify what the additive procedure "+" represents, but it is the failure of the additivity property of linearity that lies behind the oft-quoted (and misleading) slogan about emergence "The whole is more than the sum of its parts." It is also why nonlinear systems are often claimed to give rise to emergent features. We saw in section 3.1 that Mill relied on the decomposability of invariant causal influences for his account of causation, and also used the additivity of resultant forces as a core case of a homopathic law. As far as I am aware, he never appealed to linearity, but the associations among causal invariance, the failure of additivity, and emergence are clear in his approach. Linearity also captures the separability of influences on a system in this way: if all of the inputs x_i to the system are held constant except for x_j, then changes in the output of the system depend only upon changes in the value of x_j.

One other important use of linearity is the ability to take a spatial problem, divide the region into subregions, solve the equation for each subregion, and stitch the solutions together. Linear equations thus parallel the methods of analysis and decomposition. In contrast, as David Campbell puts it: "one must consider a non-linear problem *in toto*; one cannot—at least not obviously—break the problem into small subproblems and add the solutions" (1987, 219).[14]

So, here we have a glimpse of the holistic feature that is characteristic of emergent phenomena. Campbell identifies three characteristic differences between linear and nonlinear systems. The first is that they describe qualitatively different types of behavior. Linear systems are (usually) described by well-behaved functions, whereas nonlinear systems are described by chaotic or irregular functions. This difference can be seen in the transition from laminar to turbulent motion, for which the former exhibits linear behavior whereas the latter is highly nonlinear. The second characteristic is that in linear systems, a small change in parameters or a small external perturbation will lead to small changes in behavior. For nonlinear systems, such small changes can lead to very large qualitative changes in the motion. This behavior of linear systems is related to the characteristic of stability under external perturbations that is sometimes considered to be a feature of systems that do not exhibit emergent behaviors. The third characteristic is that linear systems exhibit dispersion, as in the decay of water waves moving away from a central source, whereas the stability of eddies in turbulent flow exhibits nondispersion. This can be associated with pattern emergence and the three kinds of autonomy discussed in section 4.2.

However, things are different when we have a completely integrable system. Rueger (2000a, 318) notes that a completely integrable

14. See also Strogatz (1994, 8–9).

system can be represented as one consisting of a number of independent subsystems:

> A dynamical system with N degrees of freedom . . . is said to be (completely) integrable if it has N "first integrals of motion" for all initial conditions and parameter values. Intuitively this means that if a system is integrable, we can always find a representation of it (by transforming the original description of the system) that describes a collection of N *non-interacting or uncoupled* constituent systems. . . . *Essential interaction* occurs in a system if the actually interacting components cannot be replaced by an equivalent system of non-interacting parts.

Hamiltonian systems with one degree of freedom are completely integrable and the problem of their state space trajectory can be solved analytically using initial value approaches (Campbell 1987, 221).[15] One will, thus, not encounter inferentially emergent phenomena in a completely integrable system if one takes the in-principle approach to predictability. The reference to analyticity that is used here is important, because it is the breakdown of analyticity that gives rise to emergent phenomena at critical points. The failure of the superposition principle is also far from fatal to predictability; as one example, the equation of motion for a nonlinear pendulum can be given a closed form solution for arbitrary initial conditions (Campbell op.cit., 221).

7.2.2 Complexity Theory

Complexity theory is a rich source of insights and examples of emergence, but it is often not made explicit why the examples are cases of emergence. Although the term "complexity theory" covers a wide variety of methods, one of its goals is to find general features of complex systems that hold independently of the subject matter. We can identify two rather different approaches to representing complex systems.[16]

15. Although there are Hamiltonian systems with N >1 degrees of freedom that are completely integrable, they are the exception.
16. Carlson and Doyle (2002, 2538).

In physics, the emphasis is on systems with large numbers of degrees of freedom that have strong interactions between the elements and many scales or levels, usually nonlinear, driven and dissipative. In information theory, the emphasis is on algebraic complexity and the information contained in a large sequence of data with the use of complexity measures and information as entropy.

There is no generally accepted definition of a complex system, but we can see why complex systems have often been associated with emergence, using the following definition, which is adapted from a characterization given in Prokopenko, Boschetti, and Ryan (2009):

(1) Complex systems are open systems with a flow of energy, information, or some other quantity.
(2) They contain many interacting components, the number of which is smaller than required for statistical mechanics and larger than can be dealt with by traditional analytical mathematics.
(3) The interactions between the components leads to coordinated global behavior that is exhibited in global patterns or properties of other kinds.

Typical features of complex systems then include:

1. The autonomy of the system, consisting in the absence of external or central internal controls.
2. Global order as a result of local interactions.
3. Dynamic operation. This precludes synchronic, instantaneous organization and forces us to consider diachronic emergence.
4. The system cannot be fully understood by analysis into its components.

We see reflected in condition 1 our criterion of autonomy, in conditions 2 and 4 our criterion of holism, and in condition 3 our criterion that the emergent feature must result from some other features—in this case, by a diachronic process. Perhaps also the novelty feature is associated with condition 4.

Emergence has been widely discussed within complexity theory, although not always explicitly, by adopting a combination of pattern emergence and a diachronic version of the unpredictability approach to emergence. Although the patterns involved are, simply in virtue of being patterns, nonlocal, they are not the result of a central organizing principle but, rather, result from local, often nonlinear interactions between members of a population. Examples of such patterns abound in self-organized systems. More generally, agent-based or individual-based models can provide canonical examples of complex systems, because they combine a bottom-up commitment to individualism with the dynamic emergence of higher level structures possessing the features of novelty and holism.

It is because of the existence of global patterns that an important task is to find a mode of description within which stable structural features can be located. Complexity theorists sometimes write as if we already know what emergent properties are and the problem is just to predict them (e.g., Israeli and Goldenfeld 2004), whereas the difficult task is to formulate the appropriate vocabulary in the first place. This emphasis on emergent features also often leads to the models that emphasize qualitative structural features rather than detailed quantitative predictions. This is especially notable in the emphasis on topological features of attractors and basins of attraction in the state space within dynamical systems theory, discussed in the next section.

Complex adaptive systems are an important special case of complex systems. With adaptive systems, the rules governing the behaviors of the individual agents evolve as the agents interact with their environments. Because an agent's environment includes other agents, the adaptation involves responses to the behaviors of those agents. If one takes the rules that govern an agent as being partly constitutive of that agent's identity, complex adaptive systems can give rise to cases of transformational emergence. The ability of the system to develop new types of agents is accompanied by the system itself often being open. This openness is important when self-organization occurs in some systems at equilibrium, such as the occurrence of rigidity and of superconductivity, and in others, in dissipative systems that are far from equilibrium, such as convection cells and vortices in turbulent fluids. A key element is that the decrease in entropy of a self-organized system is compensated for by exporting the entropy outside the system.

7.3 DYNAMICAL SYSTEMS

I shall now show how diachronic accounts of emergence that draw on dynamical systems theory appeal to a number of different aspects of emergence, but do not themselves constitute a single category of emergence. To do this we shall need a number of definitions from the general area of dynamical systems. In what follows, the systems are always model systems rather than real systems.

A *system* is a collection of objects among which relations hold. The relations can be of many different kinds, but a completely unstructured collection—one such that there are no relations between the objects that are explicitly included in the model—is not a system.[17] Hence, any system will possess at least one structural feature. An attribution of values to the intrinsic properties of the objects of the system, and a particular relational configuration of those objects, is a *state* of the system. The set of all possible states of the system is the *state space* of the system. Then a *dynamical system* is a system for which the states change according to explicitly specifiable rules (which may be deterministic or indeterministic).[18] What generally needs to be represented is how the state of the system changes over time, and the computational challenge is to use the representation to calculate future states from past and present states. Dynamical systems theory is, thus, the study and description of the evolution of the states of a system.

A *conservative* dynamical system is one in which there is no flow of energy across the boundary of the system. When such a flow of energy

17. There are always relations between any N objects, but if the only properties explicitly included in the model are monadic properties, the model does not describe a system.
18. More formally, a dynamical system is a triple $<M, f, T>$, where M is a manifold, *f* is a (time-dependent) diffeomorphism of M to M, and T is either a continuous or a discrete representation of time. An alternative definition is that a dynamical system is a quadruple $<X, F, \mu, \tau>$, where $<X, F, \mu>$ is a probability space and τ is a measure-preserving map from X to X.

exists, we have a *dissipative* system, which is one in which volumes in phase space contract under dynamic flows.[19] The damped pendulum is a dissipative dynamical system, whereas the free pendulum is a conservative system. Conservative systems cannot have any attracting fixed points, but they can have other kinds of fixed points.[20]

Dynamical systems theory is important for a number of reasons. The theory allows a qualitative representation of important features of the dynamics without the need to take into account the details or the nature of the system, an enormous advantage when dealing with complex systems. The representations may, thus, be applied to a wide variety of types of systems, with these applications ranging from classical mechanics to neural nets, and we see here a reflection of the autonomy from micro details that is one characteristic of emergence. Attractors in dynamical systems theory can also be seen as the basis of self-organization—in this case, spontaneously steering a system's dynamics into a predetermined end-state, a view that was suggested in the 1950s by Ross Ashby.

Alexander Rueger (2000a, 2000b) has developed a version of diachronic emergence that falls into the category of weak emergence. The approach takes higher level properties as structural properties, and the aim is to draw a distinction between merely resultant structural properties and genuinely novel structural properties: "We can say that the behavior of the system at a time is emergent with respect to the system at an earlier time if some parameter in the base has changed its value slightly during the time interval and the later behavior is 'novel' compared to the behavior of the old system and irreducible to it" (2000a, 300). A sufficient condition for novelty is that no smooth deformation transforms the set of trajectories resulting from one value of the control parameter into the set of trajectories resulting from another value of the control parameter.

19. See Strogatz (1994, 312).

20. Although attractors are usually classified as point attractors, periodic attractors, strange attractors, and so on, there is no generally agreed choice for a classification of attractors, nor a complete classification. Frequently, asymptotic behavior has to be examined in order to discover the global features of the system.

For example, the limit cycle for a damped oscillator as described in phase space requires a topologically novel description compared to that of an undamped oscillator. Although Rueger does not describe it as such, this is an example of conceptual emergence because it requires a conceptual redescription (that of a new kind of attractor) to capture the new behavior. The approach is diachronic, because the change in attractor type emerges from the change in the control parameter. Rueger's appeal to discontinuities is similar to the phase transition account of emergence, in that the topological discontinuity occurs at a critical value of the control parameter.[21]

Stephen Wolfram's classification of one-dimensional cellular automata is also relevant to dynamical systems. Cellular automata fall into four classes. In Class I, a unique homogeneous state is produced that destroys information about the initial state. Predicting the dynamic evolution of Class I cellular automata is trivial and their complexity is low. These are analogous to fixed points in phase space. In Class II, each region of the final state depends only on a finite region of the initial state. There is thus again a high level of predictability. This is analogous to limit cycles in phase space. Class III includes infinite cellular automata that produce chaotic spatiotemporal patterns from almost all possible initial states. As the computation proceeds further into the future, a region of

21. An ancestor of both Rueger's position and Batterman's approach can be found in Hans Primas's (1991) work: "Most intertheoretical relations are mathematically describable as *singular* asymptotic expansions. . . . Asymptotic expansions are never universally convergent but valid only in a very specific context. . . . The context-dependence of any description of nature is intrinsically related with the fact that we can say nothing about nature without making abstractions. . . . The abstractions we introduce into experimental sciences have also to be introduced into a theoretical description by corresponding new *contextual topologies* which as a rule induce *symmetry breakings*. . . . A mathematically rigorous deduction of emergent quantities from a basic theory requires a breaking of fundamental symmetries, either by a singular asymptotic expansion which does not converge in the topology of the primary theory, or by choosing one of the infinitely many possible, physically inequivalent representations of the primary theory. . . . As a rule, symmetry breakings induce the emergence of *hierarchical levels* with clearly separated time ranges. But it would not be correct to say that nature *is* structured hierarchically. It is our *viewpoint with the associated abstractions* which generate the higher levels of a hierarchical description" (4–5).

the final state depends on an ever-increasing region of the initial state. This is analogous to strange attractors in phase space. Class IV deals with automata that generate patterns and those patterns continuously change. An important feature of Class IV cellular automata is that they can carry out universal computations (see Wolfram 1984).[22]

22. The preceding paragraph is based on information in Prokopenko et al. (2009).

REFERENCES

Aaronson, Scott. 2003. "Book Review on *A new kind of Science* by Stephen Wolfram." *Quantum Information and Computation* 2: 410–23.

Abbott, Russ. 2006. "Emergence Explained: Abstractions: Getting Epiphenomena to do Real Work." *Complexity* 12: 13–26.

Aegerter C. M., R. Günther, and R. J. Wijngaarden. 2003. "Avalanche Dynamics, Surface Roughening, and Self-Organized Criticality: Experiments on a Three-Dimensional Pile of Rice." *Physical Review E* 67: 051306-1–051306-6.

Aharoni, Amikam. 2000. *Introduction to the Theory of Ferromagnetism*, 2nd ed. Oxford: Oxford University Press.

Alexander, Samuel. 1920. *Space, Time, and Deity: The Gifford Lectures at Glasgow 1916–1918*, vols. 1 and 2. London: Macmillan.

Anderson, P. W. 1972. "More Is Different." *Science* 177: 393–96.

Anderson, P. W. 1995. "Historical Overview of the Twentieth Century Physics." In *Twentieth Century Physics*, ed. L. M. Brown, A. Pais, and B. Pippard, 2017–32. New York: American Institute of Physics Press.

Anderson, P. W. 2005. "Some General Thoughts About Broken Symmetry." In *A Career in Theoretical Physics*, 2nd. ed., ed. P. W. Anderson, 419–30. Singapore: World Scientific.

Anthony, Louise. 1999. "Making Room for the Mental." *Philosophical Studies* 95: 37–44.

Armstrong, David. 1968. *A Materialist Theory of Mind*. New York: Humanities Press.

Armstrong, David M. 1978a. *Nominalism and Realism, Vol. I: A Theory of Universals*. Cambridge: Cambridge University Press.

Armstrong, David. 1978b. *Universals and Scientific Realism, Volume II: A Theory of Universals*. Cambridge: Cambridge University Press.

Armstrong, David. 1989. *Universals: An Opinionated Introduction*. ? : Westview Press.

Armstrong, David. 1997. *A World of States of Affairs*. Cambridge: Cambridge University Press.

Arntzenius, Frank. 2008. "Gunk, Topology, and Measure." *Oxford Studies in Metaphysics*, vol. 4. New York: Oxford University Press.

Assad, Andrew M., and Norman Packard. 1992. "Emergent Colonization in an Artificial Ecology." In *Toward a Practice of Autonomous Systems: Proceedings of the First European Conference on Artificial Life*, ed. Francisco J. Varela and Paul Bourgine, 143–52. Cambridge, MA: MIT Press.

Auyang, Sunny. 1998. *Foundations of Complex-System Theories*. Cambridge: Cambridge University Press.

Bailey, David, Peter Borwein, and Simon Plouffe. 1997. "On the Rapid Computation of Various Polylogarithmic Constants." *Mathematics of Computation* 66: 903–13.

Bak, P., C. Tang, and K. Wiesenfeld. 1987. "Self-Organized Criticality: An Explanation of 1/f Noise." *Physical Review Letters* 59: 381.

Baldwin, T. 1990. *G.E. Moore*. London: Routledge.

Batterman, Robert. 2000. "Multiple Realizability and Universality." *British Journal for the Philosophy of Science* 51: 115–45.

Batterman, Robert. 2002. *The Devil in the Details: Asymptotic Reasoning in Explanation, Reduction, and Emergence*. New York: Oxford University Press.

Bedau, Mark. 1997. "Weak Emergence." In *Philosophical Perspectives: Mind, Causation, and World*, ed. J. Tomberlin, vol. 11, 375–99. Malden: Blackwell.

Bedau, Mark. 2002. "Downward Causation and Autonomy in Weak Emergence." *Principia Revista Internacional de Epistemologica* 6: 5–50. Reprinted in *Emergence: Contemporary Readings in Philosophy and Science*, ed. Mark Bedau and Paul Humphreys, 155–88. Cambridge, MA: MIT Press.

Bedau, Mark. 2008. "Is Weak Emergence Just in the Mind?" *Minds and Machines* 18: 443–59.

Berlekamp, Elwyn R., John H. Conway, and Richard K. Guy. 2004. *Winning Ways for Your Mathematical Plays*, 2nd ed., vol. 4. Wellesley, MA: A.K.Peters/CRC Press.

Bigelow, J., and R. Pargetter. 1990. *Science and Necessity*. Cambridge: Cambridge University Press.

Binney, J., N. Dowrick, A. Fisher, and M. Newman. 1992. *The Theory of Critical Phenomena*. Oxford: Clarendon Press.

Bishop, Robert C. 2008. "Downward Causation in Fluid Convection." *Synthese* 160: 229–48.

Bishop, Robert, and Harald Atmanspacher. 2006. "Contextual Emergence in the Description of Properties." *Foundations of Physics* 36: 1753–77.

Blitz, David. 1992. *Emergent Evolution*. Dordrecht: Kluwer Academic Publishers.

Bohr, Nils. 1935. "Can Quantum-Mechanical Description of Physical Reality Be Considered Complete?" *Physical Review* 48: 696–702.

Broad, C. D. 1925. *The Mind and Its Place in Nature*. London: Routledge and Kegan Paul.

Broad, C. D. 1933. *An Examination of McTaggart's Philosophy*, vol. 1. Cambridge: Cambridge University Press.

Butterfield, Jeremy. 2011a. "Emergence, Reduction, and Supervenience: A Varied Landscape." *Foundations of Physics* 41: 920–59.

Butterfield, Jeremy. 2011b. "Less Is Different: Emergence and Reduction Reconciled." *Foundations of Physics* 41: 1065–135.

Butterfield, Jeremy. 2012. "Laws, Causation, and Dynamics at Different Levels." *Interface Focus* 2: 101–14.

Callender, Craig. 2001. "Taking Thermodynamics Too Seriously." *Studies in History and Philosophy of Science, Part B: Studies in History and Philosophy of Modern Physics* 32: 539–53.

Campbell, David K. 1987. "Nonlinear Science: From Paradigms to Practicalities." *Los Alamos Science* 15: 218–62.

Campbell, Donald. 1974. "'Downward Causation' in Hierarchically Organized Biological Systems." In *Studies in the Philosophy of Biology: Reduction and Related Problems*, ed. Francisco Ayala, 179–86. Berkeley: University of California Press.

Campbell, Donald. 1990. "Levels of Organization, Downward Causation, and the Selection-Theory Approach to Evolutionary Epistemology." In *Theories of the Evolution of Knowing*, ed. G. Greenberg and E. Tolbach, 1–17. Hillsdale, NJ: Lawrence Erlbaum.

Carlson, J. M., and John Doyle. 2002. "Complexity and Robustness." *PNAS* 99 (suppl. 1): 2538–45.

Carnap, Rudolf. 1947. *Meaning and Necessity*. Chicago: University of Chicago Press.

Carroll, Felix. 1997. *Perspectives on Structure and Mechanism in Organic Chemistry*. New York: Wiley.

Castellani, Elena. 2002. "Reductionism, Emergence, and Effective Field Theories." *Studies in History and Philosophy of Science, Part B: Studies in History and Philosophy of Modern Physics* 33: 251–67.

Chaitin, Gregory. 2005. "Epistemology as Information Theory: From Leibniz to Ω." Alan Turing Lecture on Computing and Philosophy, E-CAP'05, arXiv.org/pdf/math/0506552.pdf.

Chalmers, David. 1996. *The Conscious Mind: In Search of a Fundamental Theory*. New York: Oxford University Press.

Chalmers, David. 2006. "Strong and Weak Emergence." In *The Re-Emergence of Emergence*, ed. Philip Clayton and P. C. W. Davies, 244–56. Oxford: Oxford University Press.

Checkland, P. 1981. *Systems Thinking, Systems Practice*. Chichester: Wiley.

Chihara, Charles. 1982. "A Gödelian Thesis Regarding Mathematical Objects: Do They Exist? And Can We Perceive Them?" *Philosophical Review* 91: 211–27.

Clayton, Philip. 2004. *Mind and Emergence*. Oxford: Oxford University Press.

Colebrooke, Henry T. 1837. *Miscellaneous Essays*. London: W.H. Allen.

Comte, Auguste. 1830. *Cours de philosophie positive*. Paris: Bachelier.

Cook, Stephen A. 1971. "The Complexity of Theorem-Proving Procedures." In *Proceedings of the Third Annual ACM Symposium on the Theory of Computing*, ed. Michael A. Harrison, Ranan B. Banerji, and Jeffrey D. Ullman, 151–58. New York: Association for Computing Machinery.

Crane, Tim, and Hugh Mellor. 1990. "There Is No Question of Physicalism." *Mind* 99: 185–206.

Darley, Vincent. 1994. "Emergent Phenomena and Complexity." In *Artificial Life IV: Proceedings of the Fourth International Workshop on the Synthesis and Simulation of Living Systems*, ed. R. Brooks and P. Maes, 411–16. Cambridge, MA: MIT Press.

D'Espagnat, B. 1973. "Quantum Logic and Non-Separability." In *The Physicist's Conception of Nature*, ed. J. Mehra, 714–35. Dordrecht: D. Reidel.

Dennett, Daniel. 1991. "Real Patterns." *Journal of Philosophy* 88: 27–51.

Descartes, René. 1641. *Meditationes De Prima Philosophia*. Paris: Michel Soly.

Dill, Ken A., S. Banu Ozkan, M. Scott Shell, and Thomas R. Weikl. 2008. "The Protein Folding Problem." *Annual Review of Biophysics* 37: 289–316.

Dirac, P. A. M. 1929. "Quantum Mechanics of Many Electron Systems." In *Proceedings of the Royal Society London A* 123: 714–33.

Drexler, Eric. 1986. *Engines of Creation: The Coming Era of Nanotechnology*. New York: Anchor Books.

Dubucs, Jacques. 2006. "Unfolding Cognitive Capacities." In *Reasoning and Cognition: Interdisciplinary Conference on Reasoning Cognition*, ed. Mitsuhiro Okada, 95–101. Minato-ku, Tokyo : Keio University Press.

Dunn, J. 1990. "Relevant Predication 2: Intrinsic Properties and Internal Relations." *Philosophical Studies* 60: 177–206.

Eddington, Arthur. 1929. *The Nature of the Physical World*. Cambridge: Cambridge University Press.

Edmund, M, A. Ronald, Mose Sipper, and Mathieu S. Capcarrère. 1999. "Design, Observation, Surprise!: A Test of Emergence." *Artificial Life* 5: 225–39.

Edwards, Paul, ed. 1967. *The Encyclopedia of Philosophy*. New York: Macmillan.

Einstein, Albert, B. Podolsky, and N. Rosen. 1935. "Can Quantum-Mechanical Description of Physical Reality Be Considered Complete?" *Physical Review* 47: 777–80.

Eisberg, Robert, and Robert Resnick. 1985. *Quantum Physics of Atoms, Molecules, Solids, Nuclei, and Particles*, 2nd ed. New York: Wiley.

Emmeche, Claus, Simo Køppe, and Frederik Stjernfelt. 2000. "Levels, Emergence, and Three Versions of Downward Causation." In *Downward Causation:*

Mind, Bodies, and Matter, ed. Peter Bøgh Andersen, Claus Emmeche, Niels Ole Finnemann, and Peder Voetmann Christiansen, 13–34. Aarhus: Aarhus University Press.

Ewing, A. C. 1934. *Idealism: A Critical Survey*. London: Methuen.

Fodor, Jerry. 1974. "Special Sciences, or the Disunity of Science as a Working Hypothesis." *Synthese* 28: 97–115.

Fredkin, Edward. 1990. "An Informational Process Based on Reversible Universal Cellular Automata." *Physica D* 45: 254–70.

Friedman, Michael. 1974. "Explanation and Scientific Understanding." *Journal of Philosophy* 71: 5–19.

Ganeri, Jonardon. 2011. "Emergentisms, Ancient and Modern." *Mind* 120: 671–703.

Gendler, Tamar, and John Hawthorne. 2002. *Conceivability and Possibility*. Oxford: Oxford University Press.

Ghiselin, Michael. 1974. "A Radical Solution to the Species Problem." *Systematic Zoology* 23: 536–44.

Goodman, Nelson. 1951. *The Structure of Appearance*. Cambridge, MA: Harvard University Press.

Gu, Mile, and P. Alvarado. 2011. "Encoding Universal Computations in the Ground States of Ising Lattices." *Physical Review E* 86: 011116-1–011116-6.

Gu, Mile, Christian Weedbrook, Alvaro Perales, and Michael Nielsen. 2009. "More Really Is Different." *Physica D* 238: 835–39.

Guay, Alexandre, and Olivier Sartenaer. 2016. "A New Look at Emergence. Or When after Is Different." *European Journal for Philosophy of Science* 6: 297–322.

Haken, Hermann. 1988. *Information and Self-Organization*. Berlin: Springer.

Hare, R. M. 1952. *The Language of Morals*. Oxford: Clarendon Press.

Hartmann, Stephan. 2001. "Effective Field Theories, Reductionism and Scientific Explanation." *Studies in History and Philosophy of Science, Part B: Studies in History and Philosophy of Modern Physics* 32: 267–304.

Hegselmann, Rainer. Forthcoming. "Thomas C. Schelling and James M. Sakoda—How to Become an Unknown Pioneer?" *Journal of Artificial Societies and Social Simulations*.

Hempel, Carl. 1965. "Aspects of Scientific Explanation." In *Aspects of Scientific Explanation and Other Essays in the Philosophy of Science*, 331–496. New York: Free Press.

Hempel, Carl G., and Paul Oppenheim. 1948. "Studies in the Logic of Explanation." *Philosophy of Science* 13: 135–75.

Hendry, Robin Findlay. 2006. "Is There Downward Causation in Chemistry?" In *Philosophy of Chemistry*, ed. D. Baird et al., 173–89. Berlin: Springer.

Heylighen F. 2001. "The Science of Self-organization and Adaptivity." In *Knowledge Management, Organizational Intelligence and Learning, and Complexity*, ed. L. D. Kiel. Oxford: Eolss Publishers.

Holland, John. 1998. *Emergence: From Chaos to Order.* Reading, MA: Addison-Wesley.

Honderich, Ted, ed. 1995. *The Oxford Companion to Philosophy.* Oxford: Oxford University Press.

Horgan, Terrence. 1993. "From Supervenience to Superdupervenience: Meeting the Demands of a Material World." *Mind* 102: 555–86.

Hovda, Paul. 2008. "Quantifying Weak Emergence." *Minds and Machines* 18: 461–73.

Howard, Don. 2007. "Reduction and Emergence in the Physical Sciences: Some Lessons from the Particle Physics and Condensed Matter Debate." In *Evolution and Emergence: Systems, Organisms, Persons,* ed. Nancey Murphy and William Stoeger, 141–57. Oxford: Oxford University Press.

Hull, David. 1976. "Are Species Really Individuals?" *Systematic Zoology* 25: 174–91.

Humphreys, P. 1989. *The Chances of Explanation.* Princeton, NJ: Princeton University Press.

Humphreys, P. 1990. "A Conjecture Concerning the Ranking of the Sciences." *Topoi* 9: 157–60.

Humphreys, P. 1995. "Understanding in the Not-So-Special-Sciences." *Southern Journal of Philosophy Supplement: Proceedings of the 1995 Spindel Conference* 24: 99–114.

Humphreys, P. 1997a. "How Properties Emerge." *Philosophy of Science* 64: 1–17.

Humphreys, P. 1997b. "Emergence, Not Supervenience." *Philosophy of Science* 64: S337–S345.

Humphreys, P. 2004. *Extending Ourselves.* New York: Oxford University Press.

Humphreys, P. 2011. "Contextual Emergence." *Philosophical Research,* Chinese Academy of Social Sciences. (in Chinese).

Humphreys, P. 2013. "Scientific Ontology and Speculative Ontology." In *Scientific Metaphysics,* ed. Don Ross, James Ladyman, and Harold Kincaid, 51–78. New York: Oxford University Press.

Humphreys, P. 2014a. "The Dynamics of Emergence." Unpublished paper presented at the Emergence and Reduction workshop, IHPST Paris, October 2014.

Humphreys, P. 2014b. "Explanation as Condition Satisfaction." *Philosophy of Science* 81: 1103–16.

Humphreys, P. 2015. "More Is Different . . . Sometimes." In *Why More Is Different: Philosophical Issues in Condensed Matter Physics and Complex Systems,* ed. Brigitte Falkenburg and Margaret Morrison, 137–52. Berlin: Springer-Verlag.

Hütteman, Andreas. 2005. "Explanation, Emergence, and Quantum Entanglement." *Philosophy of Science* 72: 114–27.

Hüttemann, Andreas, and O. Terzidis. 2000. "Emergence in Physics." *International Studies in the Philosophy of Science* 14: 267–81.

Hylton, Peter. 1990. *Russell, Idealism, and the Emergence of Analytic Philosophy.* Oxford: Clarendon Press.

Imbert, Cyrille. 2006. "Why Diachronically Emergent Properties Must Also Be Salient." In *Philosophy and Complexity: Essays on Epistemology, Evolution, and Emergence,* ed. Carlos Gershenson, Diederik Aerts, and Bruce Edmonds, 99–116. Singapore: World Scientific.

Israeli, Navot, and Nigel Goldenfeld. 2004. "Computational Irreducibility and the Predictability of Complex Physical Systems." *Physical Review Letters* 92: 074105-1–074105-4.

Jaeger, H. M., Chu-heng Liu, and Sidney R. Nagel. 1989. "Relaxation at the Angle of Repose." *Physical Review Letters* 62: 40.

Jardine, Nicholas. 1980. "The Possibility of Absolutism." In *Science, Belief, and Behaviour: Essays in Honor of R. B. Braithwaite,* ed. D. H. Mellor, 23–42. Cambridge: Cambridge University Press.

Jensen, Henrik. 1998. *Self-Organized Criticality: Emergent Complex Behavior in Physical and Biological Systems.* Cambridge Lecture Note in Physics, vol.10. Cambridge: Cambridge University Press.

Johnston, Mark. 1997. "Manifest Kinds." *Journal of Philosophy* 94: 564–83.

Kim, Jaegwon. 1976. "Events as Property Exemplifications." In *Action Theory,* ed. M. Brand and D. Walton, 159–77. Dordrecht, D. Reidel. Reprinted in Jaegwon Kim, *Supervenience and Mind,* 33–52. Cambridge: Cambridge University Press, 1993.

Kim, Jaegwon. 1992. "Downward Causation." In *Emergence or Reduction? Essays on the Prospects of Nonreductive Physicalism,* ed. Ansgar Beckerman, Hans Flohr, and Jaegwon Kim, 119–38. Berlin: Walter de Gruyter.

Kim, Jaegwon. 1993a. *Supervenience and Mind.* Cambridge: Cambridge University Press.

Kim, Jaegwon. 1993b. "The Non-reductivist's Troubles with Downward Causation." In *Mental Causation,* ed. John Heil and Alfred Mele, 189–210. Oxford: Oxford University Press. Reprinted in Jaegwon Kim, *Supervenience and Mind,* chap. 17. Cambridge: Cambridge University Press, 1993.

Kim, Jaegwon. 1997. "Explanation, Prediction, and Reduction in Emergentism." *Intellectica* 25: 45–57.

Kim, Jaegwon. 1998. *Mind in a Physical World.* Cambridge, MA: MIT Press.

Kim, Jaegwon. 1999. "Making Sense of Emergence." *Philosophical Studies* 95: 3–36.

Kim, Jaegwon. 2005. *Physicalism, Or Something Near Enough.* Princeton, NJ: Princeton University Press.

Kitcher, Philip. 1989. "Explanatory Unification and the Causal Structure of the World." In *Scientific Explanation. Minnesota Studies in the Philosophy of Science,* ed. P. Kitcher and W. Salmon, 410–505. Minneapolis: University of Minnesota Press, 1989.

Klee, Robert. 1984. "Micro-Determinism and Concepts of Emergence." *Philosophy of Science* 51: 44–63.

Kronz, F., and J. Tiehen. 2002. "Emergence and Quantum Mechanics." *Philosophy of Science* 69: 324–47.

Kuhlmann, Meinard. 2014. "Explaining Financial Markets in Terms of Complex Systems." *Philosophy of Science* 81: 1117–30.

Kuhn, Thomas. 1962. *The Structure of Scientific Revolutions*. Chicago: University of Chicago Press.

Ladyman, James, Don Ross, David Spurrett, and John Collier. 2007. *Everything Must Go: Metaphysics Naturalized*. Oxford: Oxford University Press.

Laplace, Pierre Simon Marquis de. 1902. *A Philosophical Essay on Probabilities*. London: Chapman and Hall. English translation of the sixth French edition.

Laughlin, Robert. 2005. *A Different Universe: Reinventing Physics from the Bottom Down*. New York: Basic Books.

Leibniz, G. W. 1714/1989. *Monadology* (English translation). In *Philosophical Essays*, ed. Roger Ariew and Daniel Garber, 213–24. Indianapolis: Hackett.

Levine, Joseph. 1993. "On Leaving Out What It's Like." In *Consciousness*, ed. M. Davies and G. W. Humphreys, 121–36. Oxford: Blackwell.

Lewes, G. H. 1874. *Problems of Life and Mind*. Boston: J.R. Osgood.

Lewis, David. 1966. "An Argument for the Identity Theory." *Journal of Philosophy* 63: 17–25.

Lewis, David. 1986a. *Philosophical Papers*, vol. 2. Oxford: Oxford University Press.

Lewis, David. 1986b. *On the Plurality of Worlds*. Oxford: Basil Blackwell.

Lewis, David. 1986c. "Against Structural Universals." *Australasian Journal of Philosophy* 64: 25–46.

Lewis, David. 1991. *Parts of Classes*. Oxford: Blackwell.

Li, Ming, Xin Chen, Xin Li, Bin Ma, and Paul M.B. Vitanyi. 2004. "The Similarity Metric." *IEEE Transactions on Information Theory* 50: 3250–64.

Liu, Chang. 1999. "Explaining the Emergence of Cooperative Phenomena." *Philosophy of Science* 66: S92–S106.

Massimi, Michela. 2005. *Pauli's Exclusion Principle: The Origin and Validation of a Scientific Principle*. Cambridge: Cambridge University Press.

Masterton, W., and E. Slowinski. 1969. *Chemical Principles*, 2nd ed. Philadelphia: W.B. Saunders.

McLaughlin, Brian. 1992. "The Rise and Fall of British Emergentism." In *Emergence or Reduction? Essays on the prospects of Nonreductive Physicalism*, ed. Ansgar Beckerman, Hans Flohr, and Jaegwon Kim, 49–93. Berlin: Walter de Gruyter.

McLaughlin, Brian. 1997. "Emergence and Supervenience." *Intellectica* 25: 25–43.

Mill, J. S. 1843. *A System of Logic: Ratiocinative and Inductive*. London: Longmans, Green.

Mill, J. S. 1884. *An Examination of Sir William Hamilton's Philosophy*. New York: Henry Holt.

Moore, G. E. 1903. *Principia Ethica*. Cambridge: Cambridge University Press.

Moore, G. E. 1922. "External and Internal Relations." In *Philosophical Studies*, 276–309. London: Routledge and Kegan Paul.

Morgan, C. Lloyd. 1923. *Emergent Evolution*. London: Williams & Norgate.

Murphy, Nancey. 2006. *Bodies and Souls, or Spirited Bodies?* Cambridge: Cambridge University Press.

Nagel, Ernest. 1961. *The Structure of Science*. New York: Harcourt.

Nagel, Sidney. 1992. "Instabilities in a Sand Pile." *Reviews of Modern Physics* 64: 321–25.

Nelsen, Roger. 1999. *An Introduction to Copulas*. Berlin: Springer.

Newman, David. 1996. "Emergence and Strange Attractors." *Philosophy of Science* 63: 245–61.

Nietzsche, Friedrich. 1889/1968. "Expeditions of an Untimely Man." In *Twilight of the Idols*, trans. R. J. Hollingdale. London: Penguin.

O'Connor, Timothy, and Hong Yu Wong. 2005. "The Metaphysics of Emergence." *Noûs* 39: 658–78.

Parrish, Julia K., and Leah Edelstein-Keshet. 1999. "Complexity, Pattern, and Evolutionary Trade-Offs in Animal Aggregation." *Science* 284: 99–101.

Pelletier, F. J. 2000. "Did Frege Believe Frege's Principle?" *Journal of Logic, Language and Information* 10: 87–114.

Pepper, Stephen. 1926. "Emergence." *Journal of Philosophy* 23: 241–45.

Primas, Hans. 1991. "Reductionism: Palaver Without Precedent." In *The Problem of Reductionism in the Sciences*, ed. Evandro Agazzi.

Prokopenko, Mikhail, Fabio Boschetti and Alex J. Ryan. 2009. "An Information-Theoretic Primer on Complexity, Self-organization, and Emergence." *Complexity* 15: 11–28.

Psillos, Stathis. 1999. *Scientific Realism: How Science Tracks Truth*. London: Routledge.

Putnam, Hilary. 1962. "What Theories Are Not." In *Logic, Methodology, and Philosophy of Science*, ed. E. Nagel, P. Suppes, and A. Tarski, 240–51. Stanford, CA: Stanford University Press.

Putnam, Hilary. 1975. "Philosophy and Our Mental Life." In *Mind, Language, and Reality: Philosophical Papers*, ed. H. Putnam, vol. 2, 291–303. Cambridge: Cambridge University Press.

Rasmussen, Steen, and Chris Barrett. 1995. "Elements of a Theory of Simulation." In *European Conference on Artificial Life*, Lectures Notes in Computer Science, vol. 95, 515–29. Berlin: Springer-Verlag.

Rohrlich, F. 1997. "Cognitive Emergence." *Philosophy of Science* 64: S346–S358.

Ronald, E., M. Sipper, and M. Capcarrère. 1999. "Design, Observation, Surprise! A Test of Emergence." *Artificial Life* 5: 225–39.

Rorty, Richard. 1967. "Relations, Internal and External." In *Encyclopedia of Philosophy*, ed. Paul Edwards, vol. 7, 125–33. New York: MacMillan/Free Press.

Rueger, Alexander. 2000a. "Physical Emergence, Diachronic and Synchronic." *Synthese* 124: 297–322.

Rueger, Alexander. 2000b. "Robust Supervenience and Emergence." *Philosophy of Science* 67: 466–89.

Russell, Bertrand. 1910. "The Monistic Theory of Truth." In *Philosophical Essays*, 123–39. London: Longmans, Green.

Russell, Bertrand. 1912. "On the Notion of Cause." *Proceedings of the Aristotelian Society* 13 (New Series): 1–26.

Russell, Bertrand. 1914. "The Relation of Sense-Data to Physics." *Scientia* 16: 1–27. Reprinted in Bertrand Russell, *Mysticism and Logic and Other Essays*, chap. 8. New York: Longmans, Green.

Russell, Bertrand. 1918. "The Philosophy of Logical Atomism." *The Monist* 28: 495–527; 29: 32–63, 190–222, 345–80. Reprinted with minor changes in Bertrand Russell, *Logic and Knowledge: Essays 1901–1950*, edited by Robert Charles Marsh, 177–281. London: Unwin Hyman, 1956.

Russell, Bertrand. 1956. *Logic and Knowledge: Essays 1901–1950*, edited by Robert Charles Marsh. London: Unwin Hyman.

Salmon, Wesley C. 1980. "Causality: Production and Propagation." In *PSA 1980 Volume 2: Proceedings of the Biennial Meetings of the Philosophy of Science Association*, 49–69. East Lansing: Philosophy of Science Association.

Salmon, Wesley. 1984. *Scientific Explanation and the Causal Structure of the World*. Princeton, NJ: Princeton University Press.

Schelling, Thomas. 1978. *Micromotives and Macrobehavior*. New York: W.W. Norton.

Schweber, Silvan. 1993. "Physics, Community, and the Crisis in Physical Theory." *Physics Today*, November, 34–40.

Searle, John. 1980. "Minds, Brains and Programs." *Behavioral and Brain Sciences* 3: 417–57.

Searle, John. 1995. *The Construction of Social Reality*. New York: Free Press.

Sethna, James, P. 2006. *Statistical Mechanics: Entropy, Order Parameters and Complexity*. Oxford: Oxford University Press.

Sewell, G. L. 1986. *Quantum Theory of Collective Phenomena*. Oxford: Clarendon Press.

Shagrir, Oron. 2016. "An Advertisement for the Philosophy of the Computational Sciences." In *Oxford Handbook of the Philosophy of Science*, ed. Paul Humphreys, chap. 2. New York: Oxford University Press.

Shalizi, Cosma Rohilla, and James Crutchfield. 2001. "Computational Mechanisms: Pattern and Prediction, Structure and Simplicity." *Journal of Statistical Physics* 104: 817–79.

Shoemaker, Sydney. 1980. "Causality and Properties." In *Time and Cause: Essays Presented to Richard Taylor*, ed. Peter van Inwagen, 109–36. Dordrecht: D. Reidel.

Shimony, A. 1978. "Metaphysical Problems in the Foundations of Quantum Mechanics." *International Philosophical Quarterly* 8: 2–27.

Shimony, A. 1993. "Some Proposals Concerning Parts and Wholes." In *Search for a Naturalistic World View*, vol. 2, 218–27. Cambridge: Cambridge University Press.

Shoemaker, Sydney. 1980. "Causality and Properties." In *Time and Cause*, ed. P. van Inwagen, 109–35. Dordrecht: D. Reidel.

Silberstein, Michael, and John McGeever. 1999. "The Search for Ontological Emergence." *Philosophical Quarterly* 49: 201–14.

Silva, T. J., S. Schultz, and D. Weller. 1994. "Scanning Near-Field Optical Microscope for the Imaging of Magnetic Domains in Optically Opaque Materials." *Applied Physics Letters* 65: 658–60.

Sklar, Lawrence. 1993. *Physics and Chance: Philosophical Issues in the Foundations of Statistical Mechanics*. Cambridge: Cambridge University Press.

Steels, Luc. 1994. "The Artificial Life Roots of Artificial Intelligence." *Artificial Life* 1: 75–110.

Stephan, Achim. 1992. "Emergence: A Systematic View on Its Historical Facets." In *Emergence or Reduction: Essays on the Prospects of Nonreductive Physicalism*, ed. Ansgar Beckerman, Hans Flohr, and Jaegwon Kim, 25–48. Berlin: Walter de Gruyter.

Stephan, Achim. 1998. "Varieties of Emergence in Artificial and Natural Systems." *Zeitschrift für Naturforschung* 53: 639–56.

Stephan, Achim. 1999. "Varieties of Emergentism." *Evolution and Cognition* 5: 49–59.

Strogatz, Steven. 1994. *Nonlinear Dynamics and Chaos*. Reading, MA: Addison-Wesley.

Sugiyami, Yuki, Minoru Fukui, Macoto Kikuchi, Katsuya Hasebe, Akihiro Nakayama, Katsuhiro Nishinari, Shin-ichi Tadaki, and Satoshi Yukawa. 2008. "Traffic Jams Without Bottlenecks—Experimental Evidence for the Physical Mechanism of the Formation of a Jam." *New Journal of Physics* 10: 1–7.

Teller, P. 1986. "Relational Holism and Quantum Mechanics." *British Journal for the Philosophy of Science* 37: 71–81.

Teller, P. 1992. "A Contemporary Look at Emergence." In *Emergence or Reduction? Essays on the Prospects of Nonreductive Physicalism*, ed. Ansgar Beckermann, Hans Flohr, and Jaegwon Kim, 139–53. Berlin: Walter de Gruyter.

Thompson, E., and F. Varela, F. 2001. "Radical Embodiment: Neural Dynamics and Consciousness." *Trends in Cognitive Science* 5: 418–25.

Tuomela, Raimo. 1995. *The Importance of Us: A Philosophical Study of Basic Social Notions*. Stanford, CA: Stanford University Press.

Tye, Michael. 1989. *The Metaphysics of Mind*. Cambridge: Cambridge University Press.

van Cleve, James. 1985. "Three Versions of the Bundle Theory." *Philosophical Studies* 47: 95–107.

van Cleve, James. 1990. "Mind-Dust or Magic? Panpsychism versus Emergen tism." *Philosophical Perspectives* 4: 215–26.

van Dyke Parunak, H., Sven Brueckner, and Robert Savit. 2004. "Universality in Multi-Agent Systems." In *Proceedings of the Third International Joint Conference on Autonomous Agents and Multi-Agent Systems*, vol. 2, 930–37. Washington, DC: IEEE Computer Society.

Watkins, J. W. N. 1957. "Historical Explanation in the Social Sciences." *British Journal for the Philosophy of Science* 8: 104–17

Weinberg, Stephen. 1987. "Newtonianism, Reductionism, and the Art of Congressional Testimony." *Nature* 330: 433–37.

Whewell, William. 1840. *The Philosophy of the Inductive Sciences*. London: J.W. Parker.

Wimsatt, William. 1994. "The Ontology of Complex Systems." In *Canadian Journal of Philosophy, Supplementary Volume 20*, ed. Mohan Matthen and Robert Ware, 207–74. Calgary: University of Calgary Press.

Wimsatt, W. 2007. "Emergence as Non-aggregativity and the Biases of Reductionisms." In *Re-Engineering Philosophy for Limited Beings: Piecewise Approximations to Reality*. Cambridge, MA: Harvard University Press. A shorter version is in *Foundations of Science* 5 (2000): 269–87; or as "Aggregativity: Reductive Heuristics for Finding Emergence." *Philosophy of Science* 64 (1997): S372–S384.

Wolfram, Stephen. 1984. "Universality and Complexity in Cellular Automata." Physica D 10: 1–36.

Wolfram, Stephen. 2002. *A New Kind of Science*. Champaign, IL: Wolfram Media.

Wong, Hong Yu. 2006. "Emergents from Fusion." *Philosophy of Science* 73: 345–67.

Yates, David. 2009. "Emergence, Downward Causation and the Completeness of Physics." *Philosophical Quarterly* 59: 110–31.

Zurek, Wojciech. 2002. "Decoherence and the Transition from Quantum to Classical B Revisited." *Los Alamos Science* 27: 2–25.

INDEX

Note: Italicized numbers indicate definitions of terms